Your Chickens

A Kid's Guide to Raising and Showing

GAIL DAMEROW

Storey Publishing

The mission of Storey Publishing is to serve our customers
by publishing practical information that encourages personal independence
in harmony with the environment.

Edited by Lorin Driggs
Cover and text design by Carol J. Jessop
Production assistance by Wanda Harper Joyce
Cover photograph ©Positive Images/Jerry Howard
Illustrations by Elayne Sears; except for drawings on page 71 by Alison Kolesar
Technical reviews by Francine A. Bradley and John L. Skinner
Indexed by Northwind Editorial Services

The information in this book is true and complete to the best of our knowledge. All recom-
mendations are made without guarantee on the part of the author or Storey Publishing. The
author and publisher disclaim any liability in connection with the use of this information. For
additional information, please contact Storey Publishing, 210 MASS MoCA Way, North
Adams, MA 01247.

Storey books are available for special premium and promotional uses and for customized editions.
For further information, please call 1-800-793-9396.

Printed in the United States by Capital City Press
20 19 18

Library of Congress Cataloging-in-Publication Data

Damerow, Gail.
      Your chickens: a kid's guide to raising and showing / Gail Damerow
         p.   cm.

      Includes bibliographical references and indexes.
      Summary: Advice for choosing, purchasing, raising, housing, and showing all breeds of
chickens.
        ISBN-13: 978-0-88266-823-9; ISBN-10: 0-88266-823-4
          1. Chicken—Juvenile literature.   2. Chickens—Showing—Juvenile literature
[1. Chicken industry.   2. Chickens.]   I. Title
SF487.5.D35 1993
636.5—dc20   91-54655
                                                        CIP
                                                          AC

Contents

Why Chickens?

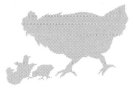

Chickens are popular for many different reasons. They are easy to raise, they don't need a lot of space, and they don't cost a lot of money to buy or to feed.

Why would anyone raise chickens?

If you have never raised livestock before, keeping chickens is a great start. All the things you learn about feeding, housing, and caring for your chickens will help you later if you decide to raise some other kind of animal.

Chickens are fun. They come when you call them. They let you pick them up and pet them. When you spend time watching your chickens, you will learn something about how all birds live and act.

Chickens are pretty. They come in all sizes, shapes, and colors. If you enter your favorites in the county fair, you may win a prize.

Chickens lay eggs. Imagine the pride you will feel when you present your family and friends with fresh eggs. You might also sell eggs to earn money to pay for your chickens' upkeep.

When you raise chickens for eggs or for meat, you will know the food you produce is safe and wholesome. You can also take pleasure in knowing that the animals producing it live under pleasant conditions.

So . . . why *not* raise chickens?

What, exactly, is involved?

Chickens need housing where they will be protected from harsh weather and other hazards. They must be brought feed and water every day.

It takes only 5 or 10 minutes to feed and water chickens and collect their eggs. But you have to do it twice a day, 7 days a week. Some days there may be other things you would rather do, but you must take time to feed and water your chickens and make sure they are safe and healthy.

You must budget your money so you can always buy feed. Even if your chickens make money for you, there will be times when no money comes in, but your chickens still have to eat.

If you raise chickens for meat, the project will be finished in 2 or 3 months. If you raise chickens for eggs, you will be responsible for them year-round.

You may become so attached to your chickens that you will be sad when it is time to trade your old hens for more efficient younger layers, or when time comes to butcher your meat birds. Things will go smoother if you remember this: *Never name a bird you plan to eat.*

Like everything else, raising chickens has bad points as well as good points. Chickens stir up dust. They produce smelly manure that must be cleaned up. Improperly managed manure attracts flies. Chickens sometimes get sick or even die.

Despite all this, raising chickens is not hard. As long as you remember that your *flock* depends on you for survival, raising chickens is quite easy.

How much does it cost to get started?

How much it costs to get started depends on the kind of chickens you want and on whether you already have some of the equipment you need. The most expensive thing you will need is a place to house your flock.

You can keep the cost down if you house your chickens in an unused tool shed or a corner of the

Flock. *A group of chickens living together.*

garage. Doing so, at least at first, lets you find out whether or not you like chickens before you spend time and money putting up special housing.

If you don't already have a fenced area, count on spending money for a fence. A fence keeps your chickens from bothering your neighbors and keeps dogs from bothering your chickens.

You will need a few other things. Some of them you can make inexpensively yourself. After reading this book, you will be ready to make a list of all the things you need. Then you can complete the start-up cost analysis on page 126.

How can I raise money to support my chickens?

Chickens offer many opportunities for earning money. You can sell their eggs to friends and neighbors. Gardeners may be interested in using the manure your flock produces as fertilizer. You might even sell feathers to handcrafters who make jewelry or fishing lures. (See Chapter 10 for more about ways to earn money from chickens.)

Since chickens offer so many different ways to earn money, you may be able to arrange a loan to finance your start-up costs. You can then pay back the loan with money you earn from your flock.

Will I need someone to help me?

The nice thing about raising chickens is that you can do most of it yourself. You may need help getting started, though, especially if you construct new housing or remodel an existing building. When you buy supplies, you may need help carrying heavy feed sacks. When other activities keep you from feeding your chickens at their usual time, you will need someone to do it for you. That someone might be a relative, a neighbor, or a friend.

You also need someone to take care of your flock

when you and your family go on vacation. It is a good idea to teach at least one person outside your family how to care for your chickens. In exchange for this favor, let your helper keep all the eggs your chickens lay while you are away.

If you plan to show your chickens, you need someone to transport you to and from the fairgrounds. Often you can get a ride with others who are participating in the same show.

Will my chickens make noise?

Male chickens are called *roosters* or *cocks*. Cocks like to crow at dawn. People once believed they crowed to scare away evil spirits that lurk in the night. A cock also crows during the day. He seems to be saying, "Look at me! I am such a handsome fellow!"

A rooster doesn't usually crow at night unless something bothers him. If you keep your chickens where dogs and other animals can't upset them and where lights don't shine on them, your rooster should not crow at night. If he does crow and the noise disturbs your family or neighbors, put your rooster in a covered pet cage at night.

If the sound of crowing during the day would cause a problem where you live, you don't need to keep a rooster at all. Your flock may contain only *hens* (female chickens). The loudest noise a hen makes occurs after she lays an egg, when she may cackle. Hens also "sing" when they are happy, but it is a soft, pleasant sound few people would object to.

Will my hens lay eggs if I don't have a rooster?

Yes. A rooster has nothing to do with whether or not a hen lays eggs. A rooster's function is to fertilize the eggs so they can develop into chicks. If you don't have a rooster, you won't be able to *hatch* the eggs your hens lay.

Hatch. The process by which a chick comes out of an egg.

If you want to raise chicks, you might borrow a rooster from another chicken keeper or buy day-old chicks to raise or fertile eggs to hatch.

What do I need to find out before I decide to raise chickens?

- First check with your family to see how they feel about chickens. You need everyone's full support, especially since you will need to ask for help from time to time.

- Check the zoning regulations for your area. Zoning laws restrict the kinds of things you are allowed to do around your house. There may be a law against keeping chickens, or the law may limit the number you can have. There may be a rule about how far you must keep your flock from the house or from the property line.

- Make sure you aren't allergic to chickens. To find out, visit the poultry show at your county fair or spend a few hours helping someone who has chickens. If you don't have any unpleasant reactions, chances are you are not allergic to chickens. If someone else in your family is allergic to chickens, maybe you can arrange to keep your flock at a friend's or neighbor's. A nearby chicken keeper may be willing to share facilities with you if you agree to help with chores.

- Let your neighbors know about your plan to raise chickens. Explain that you are going to keep their housing clean to minimize odors and flies, and that you are going to fence your chickens so they won't get into the neighbors' yards. When you let other people in on your plans, you are less likely to hear complaints later.

Where did chickens come from, anyway?

The English naturalist Charles Darwin traced chickens back tens of thousands of years to a single *breed,* the wild red jungle fowl of Southeast Asia. These chickens still inhabit the jungles of Asia. They look like today's brown Leghorns, only smaller.

Scientifically, wild jungle fowl are classified as *Gallus gallus.* All chickens are classified under the genus *Gallus,* which is the Latin word for *cock* or *rooster.*

People started keeping *Gallus gallus* in their backyards over 5,000 years ago. Chickens, even wild ones, like to stay in one place, so taming them was easy for the Asiatic people. All they did was offer a suitable place for the chickens to stay and give them something to eat.

Those early chickens didn't lay many eggs, though, and they were so scrawny they hardly made a mouthful. Chickens evolved into the useful birds they are today because, over the years, people like you selected the biggest meat birds or the best layers for breeding.

Parts of a hen and cock shown on *Gallus gallus,* the "original" chicken

The Romans called household chickens *Gallus domesticus,* a term today's scientists still use. *Domesticus* stands for *domestic,* the opposite of wild.

How did *Gallus gallus* evolve into different breeds?

Because different people look for different characteristics in a chicken, poultry breeders all over the world have developed numerous new breeds. In 1868, when Darwin took inventory of every kind of chicken in the

Chicken — Flying Dinosaur?

You can learn a lot about dinosaurs by studying chickens. Chickens and other birds are related to the dinosaur family and share many similar characteristics.

Chickens lay hard-shelled eggs and so did dinosaurs. Like chickens, dinosaurs had long, narrow heads, slender, curved necks, and short bodies. They stood upright, walked on two legs, and ran fast.

The skeleton of a chicken looks very much like the skeleton of a dinosaur. One dinosaur, *Compsognathus,* was even the same size as a chicken. Next time you have chicken for dinner, study the bones to get a pretty good idea how dinosaurs were put together.

The skeleton of a chicken looks like the skeleton of a dinosaur.

Big Chickens

The world's biggest chicken, according to *The Guinness Book of Records, 1991,* was Bruno, a rooster from Scotland who weighed 22 pounds 1 ounce.

The world's most famous big chicken was a California rooster named Weirdo. Weirdo weighed 22 pounds and went around beating up cats and dogs.

world, he found only thirteen breeds. Today we have many times that number, most of them developed during the last century.

Chickens are now the most popular domestic food animal. According to *The Guinness Book of Records,* there are six chickens for every human in the world.

A Breed for Every Need

Deciding what kind of chickens to raise is exciting because you have so many different kinds to choose from. You should have no trouble selecting a breed that suits your purpose and also looks nice to you.

Chicken Classes

Chickens come in all colors, shapes, and sizes. No one knows for sure how many different kinds exist throughout the world. New types are constantly being developed and forgotten ones rediscovered.

The *American Standard of Perfection,* put out by the American Poultry Association (APA), is a book containing descriptions and pictures of many different chickens. They are separated into large chickens and *bantam* (miniature) chickens. Each group is divided into classes.

Large chickens are classified according to their place of origin: American, Asiatic, English, Mediterranean, Continental, and Other (including Oriental). Bantams are classified according to characteristics such as whether or not they have feathers on their legs.

Each class is further broken down into breeds and varieties. The 1989 color *Standard* lists 51 breeds and 169 varieties of large chickens, plus 62 breeds and 216 varieties of bantams.

White Crested Black Polish
pullet

Bearded Black Silkie hen

Breeds

A *breed* is a group of related chickens, all having the same general size and shape, or *type*. The *Standard* shows and describes the ideal type for each breed. A chicken that is similar to the ideal for its breed is true to type or *typy*.

Some breeds have unique features other than type that make them different from other breeds. A Polish chicken, for example, has a crest of feathers on its head. The name *Polish* did not come from this chicken's place of origin — which is actually the Netherlands — but from its *poll* or feathery crown.

The kind of feathers a chicken has may also be unique. Most feathers have a smooth, satin-like surface called a *web*. The web is created by *barbicels*, tiny hooks that hold the web together. Run your fingers along a feather, top to bottom, and the barbicels let go, causing the web to separate. Rub the feather the other way, and the barbicels hook onto each other, bringing the web back together.

A feather with no barbicels does not look sleek and smooth. Instead, it looks fuzzy, like fur. The feathers of a Silkie have no barbicels, so Silkies look like birds with fur. Silkies are different from other chickens in another way. Chickens usually have white or yellow skin. Silkies have black skin.

A rooster usually has pointed feathers on his neck (*hackle*) and lower back (*saddle*). Hens have rounded feathers in those places. In a hen-feathered breed, the

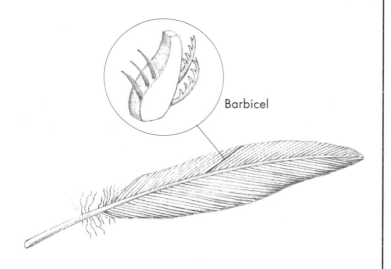

Barbicel

The smooth feather web is held together by barbicels.

Golden Sebright cock

cock has round feathers like a hen's. Sebright bantams are an example of a hen-feathered breed.

Varieties

Some breeds come in more than one color. Each different color is called a *variety*. The colors might be plain, like red, white, blue, or black. Sometimes they have patterns, like speckled, laced, or barred. Two varieties of the popular Plymouth Rock are white and barred.

Most chickens are *clean-legged*, meaning they don't have feathers growing on their legs. Some chickens have feathers growing all the way down to their toes, and they are called *feather-legged*. Some breeds, such as the Frizzle, come in a clean-legged variety and a feather-legged variety. (The Frizzle, by the way, gets its funny name from its curly feathers.)

Barred Plymouth Rock pullet

White Plymouth Rock pullet

When a chicken has a tuft of feathers growing under its chin, those feathers are called a *beard*. Some breeds, such as the Polish, come in a bearded variety and a nonbearded variety.

Instead of different colors, varieties might be separated by their style of *comb* — that reddish ornament on a chicken's head. Most breeds have a single comb with a series of zigzags like the teeth of a saw. The Sicilian Buttercup has *two* rows of points meeting at the front and back, making the comb look more like a flower. Other comb styles are pea, cushion, strawberry, and rose. Rose comb and single comb are two varieties of Leghorn.

To learn more about all the possible combinations, look through the *American Standard of Perfection*. If your town, county, or school library doesn't have a copy, ask the librarian to order one.

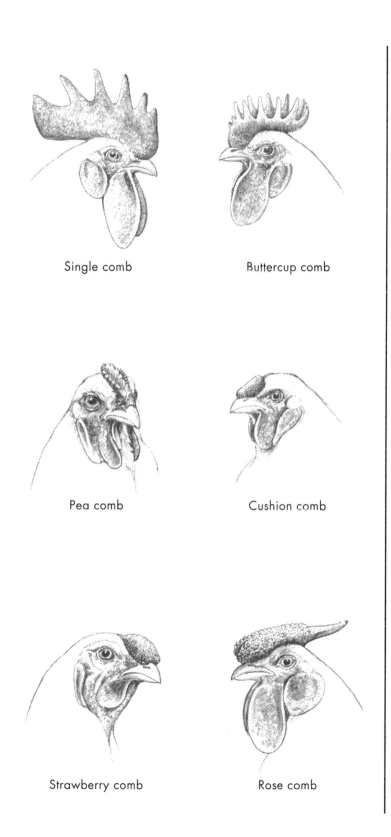

Single comb

Buttercup comb

Pea comb

Cushion comb

Strawberry comb

Rose comb

Choosing a Breed

With all of these different possibilities, how do you choose the breed that is best for you? Narrow your choices by thinking about what you want your chickens to do for you:

- Do you want lots of eggs?
- Do you want homegrown fried chicken?
- Do you want to protect an endangered breed?
- Do you want to compete at shows?
- Do you mainly want pets?

Egg Breeds

All hens lay eggs, but some lay more eggs than others. The best laying hen will give you 280 eggs, or 24 dozen, per year.

The best layers are smallish hens that lay white-shelled eggs. Since they originated near the Mediterranean Sea, they are classified as Mediterranean. Examples are Ancona, Leghorn, and Minorca. If you look at a map of the Mediterranean Sea, you will find the Spanish island of Minorca and the Italian seaport towns of Ancona and Leghorn (Livorno).

Leghorn is the breed used commercially to produce white eggs for supermarkets. Leghorns are nervous and not much fun to have. Unless you spend a lot of time taming them, they flap and squawk whenever you come near.

The most efficient layers are crosses between breeds or *strains* within a breed. The strains used to create commercial layers are often kept secret, but you can be sure that a Leghorn is involved.

Most layers produce white eggs, but some lay brown eggs. One brown-egg layer is the Golden Comet, sometimes called "the brown-egg layer that thinks like a Leghorn." Brown-egg layers are calmer than Leghorns and more fun to have.

After a few years, all hens stop laying eggs. They are called *spent* because they have spent their eggs, just like you might spend all your allowance. Unless you want to keep your old hens as pets, the best place for a spent hen is the stew pot.

Meat Breeds

Most layers are pretty scrawny because they put all their energy into making eggs instead of meat. If you want to keep chickens so you can enjoy homegrown fried chicken, raise a meat breed.

Chickens raised for meat are called *broilers, fryers,* or *roasters.* Broilers and fryers weigh about 3½ pounds and are fried or barbecued. Roasters weigh 4 to 6 pounds and are usually roasted in the oven.

Most people prefer meat birds with white feathers because they are easier to clean than dark-feathered birds. Besides being white, the best meat breeds grow plump fast.

The longer a chicken takes to get big enough to butcher, the more it eats. The more it eats, the more it costs to feed. A broiler that grows slowly costs more per pound than one that grows quickly.

Most meat breeds are in the English class. They include Australorp, Orpington, and Cornish. Of these, the most popular is Cornish, which originated in Cornwall, England.

The ideal "Cornish hen," served on special occasions, weighs exactly 1 pound *dressed,* which means ready to cook. Since preparing a chicken involves removing its feathers, you would think it made more

Spent. *No longer laying eggs.*

POULTRY PRESS

White Laced Red
Cornish hen

sense to say the bird is *undressed,* but that isn't the way it works.

The fastest-growing broilers result from a cross between Cornish and New Hampshire or White Plymouth Rock. The Rock-Cornish cross is the most popular meat bird. Those 1-pound Cornish hens are actually 4-week-old Rock-Cornish crosses.

A Rock-Cornish eats just 2 pounds of feed for each pound of weight it gains. By comparison, a *hybrid* layer eats three times as much to gain the same weight. Can you see why it wouldn't make sense to raise a laying breed for meat?

It also wouldn't make sense to keep a meat breed for eggs. Those big chickens would eat you out of house and home and reward you with too few eggs in return.

If you want both eggs and meat, you could keep a flock of layers and also raise a few fryers or roasters. Another possibility is to get a breed that produces both meat and eggs.

Hybrid. The offspring of a hen and rooster of different breeds.

Rose Comb Rhode Island Red cockerel

Dual-Purpose Breeds

A chicken kept for both meat and eggs is called a *dual-purpose* chicken. *Dual* means *two*. A dual-purpose chicken doesn't lay as well as a laying breed and isn't as fast-growing as a meat breed. But it lays better than a meat chicken and it grows plumper faster than a laying breed.

Most dual-purpose chickens are classified as American because they originated in this country. All American breeds lay brown-shelled eggs. American breeds have familiar names like Rhode Island Red, Plymouth Rock, Delaware, and New Hampshire. (You may hear New Hampshires called "New Hampshire Reds," but that isn't correct.)

Some hybrids make good dual-purpose chickens. One is the Black Sex Link, a cross between a Rhode

Island Red rooster and a Barred Plymouth Rock hen. Another is the Red Sex Link, a cross between a Rhode Island cock and a White Leghorn hen. Red Sex Links lay better than Black Sex Links, but their eggs are smaller and dressed birds weigh nearly 1 pound less. (See page 25 for an explanation of *sex link.*)

New Hampshire cockerel

Endangered Breeds

Many dual-purpose breeds, once commonly found in backyard flocks, are now endangered. These classic chickens, like their wild cousins, still know how to forage for food. Furthermore, they are fairly resistant to disease and they are able to survive harsh conditions.

The American Minor Breeds Conservancy (AMBC) lists twelve breeds and varieties as being in particular danger of becoming extinct. They are: Ancona, Barred Plymouth Rock, Black Australorp, Black Jersey Giant, Black Minorca, Brown Leghorn, Delaware, Dominique, New Hampshire, Rhode Island Red, White Jersey Giant, and White Wyandotte.

The Dominique, sometimes incorrectly called "Dominecker," is the oldest American breed. A few years ago it almost disappeared. Now it is coming back, thanks to chicken lovers who care.

If you promise to make a long-term commitment to an endangered breed, you can join the AMBC's Rare Breed Poultry Conservation Project. The address is listed at the back of this book. The AMBC specializes in dual-purpose breeds.

The Society for the Preservation of Poultry Antiquities (SPPA) specializes in exhibition chickens. The SPPA lists over 150 endangered breeds and varieties. To get a copy of that list, write the SPPA at the address given at the back of this book.

Mandy Halten and a Dominique hen raised at Wandak Farms as part of the AMBC's Rare Breed Poultry Conservation Project.

Exhibition Breeds

If you wish to show your chickens, first decide what kind of shows you want to enter. Some youth groups sponsor shows for production chickens, either layers or broilers. Many groups sponsor shows strictly for exhibition chickens.

Exhibition chickens are bred for their beauty rather than their production abilities. Some of the same breeds kept for meat and eggs are also popular for exhibition. Even though the *breed* is the same, the *strain* is different.

A strain is a related group of birds bred with emphasis on specific traits. Famous strains are identified by their breeders' names. Halbach White Plymouth Rocks, for example, were bred by H. H. Halbach of Waterford,

Wisconsin. If you visit a lot of shows, you may learn to recognize a Halbach Rock when you see one.

All the chickens in one strain are very much alike. In an exhibition strain, they are alike in appearance. In a production strain, they are alike in their laying or meat-producing efficiency.

Production strains are used by commercial chicken operations for eggs or meat. Many are hybrids, but even the pure breeds are not necessarily true to type. Exhibition strains are usually more typy but less efficient at producing eggs or meat.

Even among the exhibition breeds, some lay better than others. Exhibition Leghorns, for example, lay better than most other show breeds, even though they don't lay as well as production Leghorns. Other Mediterranean breeds that are popular for show and are also pretty good layers are Blue Andalusian, Sicilian Buttercup, and White Faced Spanish.

In the meat category, among exhibition breeds Cornish is a good choice. Cochin is another. A Cochin looks like a basketball with feathers. Both come in varieties other than white.

Black Cochin cock

Bantams

Some shows have a special division for bantams. Other shows are for bantams only. Bantams are miniature chickens weighing only 1 or 2 pounds. Bantams lay smaller eggs than large chickens. Three bantam eggs are roughly equal to two regular eggs.

Some bantams are small versions of bigger breeds, but one-fifth to one-fourth the size. Examples are Cochin, Cornish, Dominique, Leghorn, New Hampshire,

White Leghorn Bantam cock

Chicken Manners

Chickens don't have very good manners. Unless you keep your chicken in a pet carrier, bringing one into the house or car is not a good idea.

Polish, and Rhode Island Red. Other bantams come only in the miniature version. They include Antwerp Belgian, Japanese, Sebright, Silkie, and many more.

The American Bantam Association (ABA) publishes a book, called *Bantam Standard,* describing all the recognized bantam breeds. Many are not listed in the *American Standard of Perfection.*

Chickens as Pets

Any chicken makes a good pet, provided you take time to get acquainted. Flighty breeds like the Mediterraneans require more time to tame than calmer breeds like the Americans.

Chickens learn to come when you call, eat out of your hand, and sit on your shoulder while you walk around. They even enjoy watching television and going for rides.

Bantam breeds are popular as pets because they need less room and they eat less than larger breeds. The shrill crowing of a banty cock, however, is more likely to irritate neighbors than the low-pitched crowing of a larger rooster.

If you are looking for an inexpensive pet, consider a common barnyard chicken or *barny*. A barny — which may be large, bantam, or in between — is the chicken equivalent of a mongrel dog or an alley cat. Barnies can't be shown and they don't specialize in producing meat or eggs, but they are just as much fun to have as any other chicken.

If you won't have lots of time to spend with your chickens, but you have plenty of space to let them roam, you might like Old English Game, large or bantam. Old English still have the instincts of wild jungle fowl and get along quite nicely with little care.

Poultry Press

White Old English Game
Bantam cock

Buying Chickens

After settling on a breed, you have to decide whether to purchase newly hatched chicks or grown chickens. Starting with chicks gives you a chance to make friends with your birds right from the start. If you keep them warm and dry and make sure they have plenty to eat and drink, they should grow up to be healthy, happy chickens.

When you start with chicks, though, you can't tell exactly what they will look like when they grow big. A chicken's looks are important if you plan to show, since you may or may not end up with a prize winner.

Some chicks can be purchased *sexed,* which means you know when you buy them how many are *cockerels* (young cocks) and how many are *pullets* (young hens). You can also purchase chicks *straight run* or *as-hatched,* meaning they are not sexed. About half will be cockerels and half pullets. Straight run chicks are cheaper than sexed chicks.

When you start with grown-up chickens, you can easily tell which are the roosters and which are the hens. If you plan to show, you can see how true the chickens are to type. If you want chickens for meat or eggs, you won't have to wait weeks or even months, as you do when you start with chicks.

On the other hand, older birds are more likely than newly hatched chicks to be diseased, and treating sick chickens is no fun. Besides, some sellers are not very nice and may sell you spent hens. To make sure you are buying a young chicken, look for legs that are smooth and clean and a breast bone that is soft and flexible.

What to Look For

Chickens are different from other livestock because they don't come with registration papers or any other records. The main thing to look for is good health.

When you buy chicks, make sure they are bright-eyed and perky. If they come by mail, open the box in front of the mail carrier. If anything is wrong, the carrier will help you file a claim to get either new chicks or your money back.

When you buy grown chickens, their feathers should be smooth and shiny, not dull or ruffled. Their eyes should be bright, not watery or sunken. Their *shanks* should be smooth and clean, not rough and dirty. Their combs should be full and bright in color, not shrunken and dull.

Look under each chicken's wing and around the *vent,* the opening under its tail. There shouldn't be any insects crawling on its skin or any clusters of tiny insect eggs clinging to the base of its feathers.

Listen for coughing or sneezing. If some of the chickens in a flock have a cold, chances are they will all catch it. When you visit a seller, whistle as you come near the flock. The chickens will stop whatever they are doing to listen and you can easily hear any coughing or sneezing.

Shank. The part of a chicken's leg between the claw and the first joint.

Finding a Seller

Where you buy your chickens depends on what kind you want.

Production breeds. If you are looking for chicks of a meat or egg breed, buy them from a commercial *hatchery*. A hatchery is a place where eggs are brought for hatching. If there is no hatchery nearby, look for one that ships by mail. You should be able to get a list of hatcheries from your county Extension Service agent.

Another place to buy chicks of a production breed is at a feed store. Chicks sold at feed stores usually come from a hatchery, but occasionally they are hatched by a local person who enjoys raising chickens.

Sometimes you can find a nearby seller through the classified section of your local newspaper. Visiting a seller is fun because you get to ask questions. One question to be sure to ask is whether or not it is okay for you to telephone the seller if you have more questions later.

Visiting the seller lets you see what the chickens look like and how they live. You might decide you don't want those chickens, after all. Maybe they don't look exactly like you expected or they live in dirty, unhealthy conditions. Don't feel embarrassed if you change your mind after you get there. When you are spending money, you are entitled to get what you want.

Show breeds. If you want show chickens, the best place to find them is at a show. Many people enter shows to advertise chickens for sale. Even if you can't buy chickens at the show (some shows prohibit sales), you will meet people who have chickens for sale at home.

Several catalogs sell chickens by mail. If you order from a catalog, your birds will likely not be good enough to show, but they will make fine pets.

Prices

Production chickens are usually less expensive than exhibition breeds, and barnyard chickens are the least expensive of all. If you join a youth group project and everyone in your group decides to work with the same breed, buy all the chickens from one place and you are likely to get a discount.

In some areas, you can join a *chick-chain*. In spring you will get a number of chicks for free, but you must return a few grown chickens in the fall. It is a little like taking out a bank loan and paying interest. Call your county Extension Service agent to find out if your area has a chick-chain. If not, ask what you can do to help organize one.

How Many Chickens?

Take care not to get too many chickens too fast. Otherwise you run the risk of crowding them, causing them to fight or get sick.

Most of the chickens in your flock should be hens. If you have too many roosters, they will fight.

If you are interested in egg production, remember that hens don't lay at the same rate all year. Sometimes you will have more eggs than you can use. Other times you will have too few. On average, three hens will give you two eggs each day. Let's say you want six eggs per day. To figure out how many hens you need, divide 6 (eggs) by 2 (eggs) and multiply by 3 (hens):

$$6 \div 2 \times 3 = 9$$

To get six eggs per day you would need at least nine hens.

If you are getting chickens for show or as pets, a trio makes a nice little family group. A trio consists of one cock and two hens.

Cocks and Hens

In most breeds, newly hatched chicks all look alike except to an expert. If you buy *sexed* chicks, the cockerels will be sorted out, but you have to trust the seller to sort them right. Even an expert occasionally makes mistakes.

For most hybrids, chicks can be sexed by color or by feather growth. In color-sexed chicks, cockerels are a different color from pullets. In feather-sexed chicks, cockerels develop wing feathers earlier than pullets.

Chicks that can be color-sexed or feather-sexed are called *sex linked*. Black Sex Link, Red Sex Link, and Golden Comet are examples of sex-linked hybrids.

As chicks grow, you can more easily tell at a glance which are cocks and which are hens. The comb on a rooster's head and the *wattles* dangling under its chin are usually larger and brighter than the comb and wattles of a hen.

A cock has *spurs* protruding from his legs. The older a cock gets, the longer his spurs grow. Most hens have tiny spurs or just little round knobs in place of spurs. (Game hens are one exception — they may have spurs as long as 1½ inches.)

In some breeds, the cock has feathers of a different color from the hen's. In most breeds, the rooster's hackle and saddle feathers are pointed, while a hen's feathers are rounded. A cock usually has long, sweeping tail feathers, called *sickles*. And, of course, if you listen long enough, the rooster will crow. (The parts of the hen and cock are labeled on the illustration of *Gallus gallus* on page 6.)

POULTRY PRESS

POULTRY PRESS

Black Tail White
Japanese hen (top)
and cock (bottom)

How to Think Like a Chicken

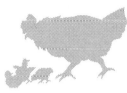

People often use expressions based on the way chickens act to describe the way people act.

- If you call someone "a chicken," you mean the person is afraid. Chickens are easily frightened by things they aren't familiar with or they don't understand.

- When you say a person has "something to crow about," you mean that person is as proud as a rooster seems to be when it crows. Someone who is "cocky" or "cock-sure" is as confident as a strutting rooster.

- A person who "gets his hackles up" — like a rooster ready to fight — is angry. So is someone who is "as mad as a wet hen."

- A "mother hen" is a person who constantly fusses over you the way a hen clucks over her newly hatched chicks.

- If your teacher describes your writing as "chicken scratching," it means your handwriting resembles the marks a chicken makes when it scratches in dirt.

- If a friend complains about an allowance being "chicken feed," you know it can't be much.

Peck Order

Chickens are social animals. They are happiest when they are with others of their own kind. In every group of chickens, one emerges as the leader.

Scientists call this "establishing the *peck order*." The peck order reduces stress by ensuring that every chicken knows how to relate to every other chicken. The leader of the flock is at the top of the peck order and is like the boss of all the other chickens. The rest are the boss of some chickens and bossed by others in the flock.

Chicks start establishing their place in the peck order when they are only 3 weeks old. If you raise baby chicks, suddenly you will notice more fighting than usual. After a couple of weeks, most of the fighting will stop. The peck order has been established. If you bring in a new chicken, it has to find its place in the peck order.

How to Act Around Your Chickens

- Whenever you approach your chicken house, whistle or sing so you won't startle your chickens. When a chicken shakes its head from side to side, you have frightened it by moving too fast or being too loud.

- The more time you spend with your chickens, the less easily frightened they will be and the friendlier they will become. Spend at least 5 minutes a day with your chickens. Walk slowly among them. Talk gently or sing to them. Pick them up and pet them.

- Your chickens will soon learn to recognize your voice and come when you say, "Chick, chick, chick." When they come running, reward them with a handful of grain or a broken up slice of bread.

Playing Chicken

The expressions "playing chicken" and "don't be such a chicken" come from peck-order squabbles. When two evenly matched birds try to decide which one is higher in rank, they may face each other and remain motionless for several minutes. The one who turns away first loses.

Most of the time, roosters are higher in rank than hens. If a hen is higher in rank than a cock, it is likely because the cock is either young and inexperienced or old and weak. That's where the expression "hen-pecked" comes from.

To find out the peck order between two of your chickens, keep all the other chickens away and offer the two something tasty to eat. The higher-ranking chicken will peck or threaten the lower-ranking one.

Another way to find out the peck order between two chickens is to pick one up and take it near the other, head first. Move the chicken in small circular motions, so it looks like it is trying to peck the other one. If the other chicken is lower in rank, it will turn away. If it is higher in rank, it will try to peck back.

Your chickens will learn to come when you call.

Pecking

Most of the time chickens peck to get something to eat. Chickens peck a little here and a little there, never eating much at one time, but eating all day long.

If you see a chicken rubbing its beak on the ground, you may think it is trying to wipe something off. But it is really sharpening its beak, just like you might sharpen a pocket knife.

When you feed your chickens store-bought rations, their beaks don't have much work to do. But if you let your chickens scratch in the yard or on the lawn, they use their beaks to bite off blades of grass and to tear apart worms or bugs.

A chicken uses its beak not just for eating but also as a weapon to attack an enemy or to defend itself. It is not a good idea to grab a chicken by surprise because it may peck you. It may even poke a hole in your skin with its sharp beak.

When you reach under a hen in a nest to see if she has eggs beneath her, she will likely peck your hand. Usually it is a gentle peck that doesn't hurt. But if the hen feels the need to protect those eggs until they hatch, she might peck pretty hard to keep you from taking them.

Chicks learn about the world by pecking. They peck at shiny things, looking for water. Until they learn better, they try to peck their mother's shiny eyes. Sometimes they peck each other's toes, or their own toes, mistaking them for worms. If you wear sandals, your chickens may investigate your toes by pecking at them.

A rooster who thinks you are upsetting one of his hens may peck you. A mean rooster pecks just for fun.

Mean Roosters

Most roosters are not mean, but most will defend their flock from a threat. The trouble is, roosters can't always tell when a threat is real. Something as simple as floppy shoelaces or baggy pants may seem like a real threat to your rooster.

A rooster may learn to be mean if he gets teased by a person or by another animal. Little children sometimes poke sticks at chickens, and dogs enjoy chasing them. Can you imagine how scared you would feel if a big giant poked a stick in your face or a monster the size of an elephant came running and barking at you?

Symbol of Valor

Don't get the idea that chickens are cowardly. In many ways, they are among the most courageous animals. In Roman times the chicken was a symbol of valor. Generals extolled the chicken's virtues to warriors before going into battle.

The best way to keep a rooster from getting mean is by acting calm around your chickens and trying not to frighten them. Keep dogs and little children away or teach them how to behave properly.

Sometimes a rooster gets the funny idea you are a big chicken and he will try to outrank you in the peck order. He may move toward you sideways, with his head down, pretending he is looking at something on the ground. His ability to look someplace else while he is really watching you is where the term "cock-eyed" comes from. The rooster may not actually attack you. He might just threaten you by rushing at you or bumping against your leg. Try to make friends with him. Pick him up and rub his wattles. Roosters like that.

When a rooster does attack, pay attention to what you are wearing or carrying. Maybe he doesn't like your rubber boots, or your flip-flop sandals, or your bare legs, or the way the feed bucket swings in your hand. Next time, try something different and see if he still gets upset.

If the rooster "gets his hackles up" — raises the feathers on his neck to make himself look big and fierce — he is serious about show-ing you who is boss. Watch out! He might fly at your leg and dig in with his claws and spurs.

You may run across a rooster who stays mean no matter what you do. Get rid of him. Like the class bully, mean roosters are no fun.

Introducing a New Chicken

When two chickens meet for the first time, right away they want to establish peck order. But the farther a chicken is from home, the more timid it becomes. The

A rooster "gets his hackles up" to make himself look big and fierce.

Hypnotizing Chickens

Next time you are looking for something interesting to do, hypnotize a chicken. Place the chicken on its back and gently rub both sides of its breastbone with your fingers. You will know the chicken is hypnotized if it doesn't move when you let go and walk away. It will stay hypnotized for a short while — a few seconds or a minute or so.

chickens already living at your place therefore have an advantage over a new chicken. The newcomer will get beaten up if you simply turn it loose in your flock.

Instead, keep the new chicken apart from the flock for a while. After giving the new chicken time to get used to you, start bringing in your other chickens, one at a time. Introduce the lowest-ranking ones first. They are usually the youngest or the oldest birds.

After the new chicken has met all your other chickens one by one, put her with the whole flock. She should get along fine because she has already established her place in the peck order. She will probably rank somewhere in the middle.

Chickens and Other Animals

Chickens get along well with other pets and livestock. You can keep them with sheep, goats, or cows. It is not a good idea, though, to keep them with ducks or geese. Ducks and geese like to play in water, but chickens like to be dry. Don't keep chickens with pigs or turkeys, either, because of the possibility of spreading diseases.

If you have dogs or cats, your chickens will get along with them, provided your pets are properly trained. The best way to train a pet is to introduce it to chickens when it is still a puppy or kitten. To a small animal, chickens look big and scary, and the puppy or kitten will learn to stay away.

Discourage your dog from playing with your chickens. Dogs play rough and can hurt a chicken without meaning to. Make your dog understand that these are *your* chickens and you expect the dog to help you guard them.

Cats are more independent than dogs and aren't as easy to train. Cats like to stalk baby chicks. When a mother hen sees a cat coming, she will squawk and fly at the cat, chasing it away. If you raise baby chicks without a hen to protect them, make sure the cat can't get them.

For more about learning to understand your chickens, see Chicken Sounds in the Glossary on pages 145–146.

Telling Chickens Apart

If your chickens look so much alike that you can't tell them apart, give each one a leg band with a number on it. Numbered leg bands are sold through poultry supply catalogs and at some feed stores. Get the right size band for your chickens.

- A number 7 band ($^{7}/_{16}$") fits most bantams.
- A number 9 ($^{9}/_{16}$") fits hens of the light breeds such as Leghorn and Ancona.
- A number 11 ($^{11}/_{16}$") fits cocks of the light breeds and most dual-purpose chickens such as Wyandotte and Plymouth Rock.
- A number 12 ($^{12}/_{16}$") fits cocks of the heavy breeds such as Jersey Giant and Cornish.

Whenever you show a chicken, you must identify it by its band number. Even if you don't show, numbered bands give you an easy way to keep track of individual chickens in your flock.

Catching a Chicken

The easiest way to catch a chicken is to go out at night, when the chicken is asleep. Reach beneath the chicken and grasp it by the legs. Cradle the chicken in one arm while you hold its legs with the other hand.

When you carry more than one chicken, turn them upside down and use their legs as handles. Hold on to both legs, or a chicken may churn around like an eggbeater until it gets away.

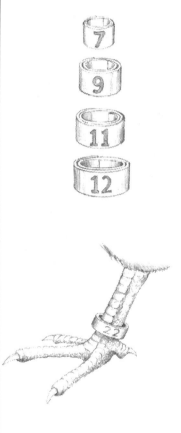

Leg bands help you identify individual chickens.

You may occasionally need to catch a chicken during the day. If your chickens are tame, walk right up to one and pick it up. If your chickens aren't tame, they will run away when you try to catch them. One way to catch a running chicken is to trap it in a corner, which is easier if you have help. Another way to catch a running chicken is to use a crook to snare its leg. Buy a crook from a poultry supply house or make one from heavy wire, such as 10 gauge, that doesn't easily bend.

Once you start to catch a chicken, don't give up. If you let it get away, it will always try to evade you in the future.

After you have caught the chicken, hold it for a moment until it calms down. Stroke its neck and wattles to let it know you meant no harm.

The best way to carry a chicken is to cradle it in one arm while you hold its legs with the other hand.

One way to catch a running chicken is to use a crook.

Chickens Come Home to Roost

Chickens need a clean place to get out of the sun and rain, and a place to sleep. They also need protection from their natural enemies.

Coming Home to Roost

You may have heard the expression, "Chickens always come home to roost." Chickens are territorial animals. They roam during the day looking for food, but at night they like to return to a familiar place. Wild chickens in the jungle sleep in a tree at night. As long as the tree remains safe and doesn't get too crowded, the chickens go back, night after night.

Thousands of years ago the people of Southeast Asia took advantage of the chickens' habit of coming home to roost. They built a little house and encouraged wild chickens to live there. A horizontal piece of wood, resembling the branch of a tree, was placed inside for the chickens to perch on at night. Nests were built inside the coop so hens wouldn't stray far to lay their eggs. The people could then easily find the eggs, and the eggs were protected from egg-eating predators such as snakes, skunks, and wild birds.

Wild chickens sleep in trees at night.

Overcrowding

Crowding leads to stress that causes chickens to eat each other's feathers or to pick at and injure one another — a nasty habit called *cannibalism*.

Your Chicken Coop

If you live in a northern climate, you must have an insulated hen house to keep your chickens warm in winter. If temperatures dip below freezing, a heater will prevent your cock's comb and wattles from freezing. Chickens can stand quite a bit of cold weather, though, if you make sure their housing stays dry inside.

Building a chicken coop is no more difficult than building a playhouse or a tree house. You can make it plain and simple or cute and fancy. You might buy a ready-made tool shed at the hardware store, or you might design a coop and build it from scratch. Get your family and friends to help and the project will go much faster.

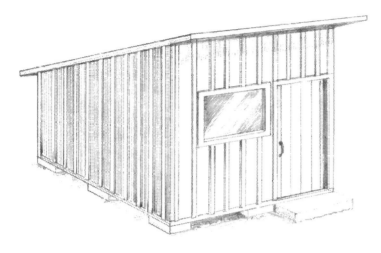

A simple hen house

Coop Location and Design

The best place for a chicken coop is on a hill or slope where puddles don't collect when it rains. If the land where you live is flat, avoid low spots where moisture collects. Puddles and muddy ground cause chickens to get sick.

Since chickens generate a certain amount of odor, dust, and noise, don't forget to check with your local zoning board to find out if laws specify how far your chickens must be from your house or from your property line.

The more room your chickens have, the happier and healthier they will be.

A coop measuring 8 feet by 12 feet is big enough for thirty regular-sized chickens or fifty bantams. For easy cleaning, the coop should be tall enough so you can stand up while you are inside. Bumping your head while cleaning house is no fun. Allow a little extra height, in case you grow taller.

The coop should have a door you can close and latch at night to protect your flock from chicken-eating animals. Dogs, coyotes, raccoons, and other

Chicken Coops from Existing Structures

■ If you own a play-house you have outgrown, convert it to a hen house by adding a perch and some nests.

■ Divide off a portion of the garage and keep your chickens there. Chickens stir up dust as they move around though. If the garage is used for other things, such as storing a shiny car, house your chickens there only temporarily while you find a better place for them.

■ An unused tool shed or other building can be used as a chicken coop.

■ If your climate is mild, a camper shell from a pickup truck, raised up on concrete blocks, makes a good hen house.

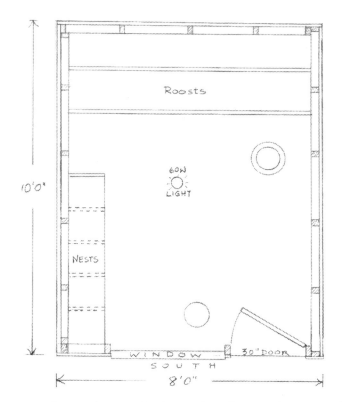

Hen house design. On the north wall, provide screened openings for summer ventilation with shutters for winter protection.

predators might otherwise wander into your coop looking for a snack.

Each evening at dusk, shut the door and make sure it is latched tight. Your chickens will already be inside, having gone in when it got dark to pick out a good place on the perch to spend the night. Provide a few screened openings, so the chicken house remains well-ventilated even when the door is shut.

The Perch

An ideal chicken perch is about 2 inches in diameter. If you have bantams, a 1-inch-diameter perch is big enough.

Dominiques and their hen house perches

Two things that do not make good perches are plastic pipe and metal pipe. Both are too smooth for a chicken to grasp firmly.

Make a nice perch from a straight tree branch, from an old ladder or a fence rail, or from anything else the right size and strong enough to hold the weight of all your chickens. If you use new lumber, round off the corners so your chickens can wrap their toes around it.

Allow 8 inches of perching space for each chicken. If you have twelve chickens, for example, you will need a perch that is 96 inches long (8 x 12 = 96).

If your coop isn't big enough to have one long perch, use two or more. Space the perches 18 inches apart and keep them at least 18 inches from the wall, so your chickens have room to get comfortable.

Place the lowest perch 2 feet off the ground. If you have two perches, make the second one 12 inches higher than the first one. Your chickens will hop onto the first perch and use it as a step to get up to the second one. Attach the perches securely so they don't turn (if they are round) or sag.

Nests

When a hen gets ready to lay an egg, she looks for a private place where her instincts tell her the egg will be safe. If you don't have nests inside your coop, your hens may lay their eggs where you can't find them or lay them on the floor.

An egg laid on the floor can become soiled. Besides, one of your chickens might step on the egg or peck at it and break it. When a chicken accidentally finds out how good eggs taste, it may start eating them on purpose. Your other chickens may learn to become egg-eaters, too, and you will have a hard time getting them to stop.

Nests keep eggs clean and safe. An egg or two left in each nest will encourage your pullets to lay there. Since real eggs would rot if you left them in the nest, use plastic or wooden ones. You can find them in poultry supply catalogs, at hobby shops, and in stores around Easter time.

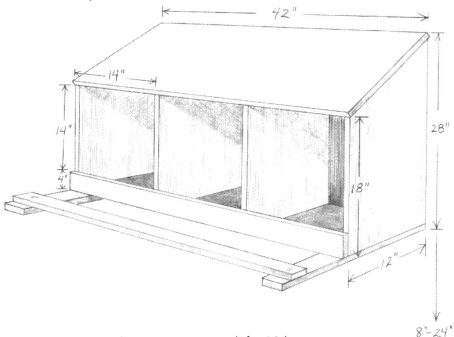

These three nests are enough for 12 hens.

Leaving eggs in your nests *won't* make your hens lay better, as some people believe. The only purpose of a nest egg is to encourage a hen to lay in the nest, instead of someplace else.

Allow one nest for every four hens. You can buy ready-made nests from a store or catalog or you can make them yourself.

A reasonable size for a nest is 14" wide by 14" high by 12" deep. Place the nests on the ground until your pullets get used to laying in them, then firmly attach the nests to the wall 1½ to 2 feet off the ground.

Put a perch in front of the nests so a hen can look inside before entering. If a nest is already occupied, she will wander down the perch until she finds a vacant one.

Nail a 4-inch-wide board across the front of the nests to keep the eggs from rolling out and to hold nesting material in.

Litter

Nesting material, also called *bedding* or *litter,* keeps eggs clean. It may be wood shavings, wood chips, sawdust, rice hulls, peanut hulls, ground-up straw, ground-up corncobs, shredded paper, or any other soft, absorbent material. Place 3 to 4 inches in each nest.

Also place a thick layer of litter on the floor of your chicken coop to keep your chickens clean and healthy. If you don't use litter, manure collects on the floor, smells bad, and attracts flies. Your chickens will scratch in the litter, stirring in manure that collects on the surface. Start with a layer at least 4 inches deep. If the litter around the doorway or under the perch gets packed down, break it up with a shovel and rake. If any litter gets wet from a leaky waterer or roof, remove the wet patch with a shovel and replace it with fresh dry litter. Be sure to fix the leak so it won't happen again.

Your chicken coop should never smell like manure or have the pungent odor of household ammonia. To keep it smelling fresh, keep the litter dry and frequently spread fresh litter on top of the old litter.

Fencing for Chicks

If you plan to keep newly hatched chicks inside the fence, they will slip through 1-inch openings. Get a roll of 12-inch-wide *aviary netting,* which looks just like chicken wire but has ½-inch openings. Tie the aviary netting tightly along the bottom of your fence.

Additional Equipment

Besides perches, nests, and litter, you will need containers for feed and water. These are described in the next chapter.

If you are raising layers, you need electricity so your hens will have 14 hours of light during winter when the days are short. Ask someone who is experienced with electricity to help you with the wiring. It is not safe to run an extension cord from the house.

A Chicken Fence

Chickens enjoy being in sunlight and fresh air. A fence keeps them from straying and protects them from dogs and other predators.

The best kind of fence for chickens is a woven fence with small openings. The most common fence material for chickens is *chicken wire,* also called *poultry netting.* It has 1-inch-wide openings woven in a honeycomb pattern.

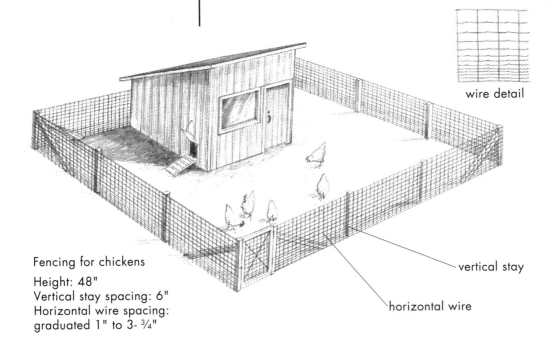

wire detail

Fencing for chickens
Height: 48"
Vertical stay spacing: 6"
Horizontal wire spacing: graduated 1" to 3-¾"

vertical stay

horizontal wire

Another kind of fence you can use with chickens is yard-and-garden fence with 1-inch spaces toward the bottom and wider spaces toward the top. The small openings at the bottom keep chickens from slipping out and keep small predators from getting in.

Your fence should be at least 4 feet high. It may need to be higher if you keep a lightweight breed that likes to fly. Bantams and young chickens of all breeds are especially fond of flying.

Pull the fence material tight and attach it to firm posts that don't wobble. To keep your chickens from sneaking out and to keep predators from crawling in, place a board on edge on the ground all along the fence and staple the bottom of the fence to it.

To further protect your chickens from predators, string electrified wire along the top and the outside bottom of your fence. Before planning and installing an electric fence, get advice from someone who uses electric fencing or from your county Extension Service agent.

Space Requirements for Chickens

Coop Size	Perch	Nests
8'x12' for 30 regular chickens or 50 bantams	8" per chicken	1 for every 4 hens

Cleaning Your Coop

If you maintain your coop properly, you should not have to clean it more than once a year. If you let it get wet, dirty, and smelly, you will have to clean it more often. You will have to clean it thoroughly, too, if your flock has had a health problem.

The best time to clean your coop is on a warm, dry spring day. Wear a dust mask so you won't breathe in the fine dust that gets stirred up. Inexpensive dust masks are available at paint and hardware stores.

First thing in the morning on cleaning day, remove all moveable feeders, waterers, perches, and nests. Then, shovel out the used litter. With a hoe, scrape caked manure from the perches, walls, and nests. Use an old broom to brush dust and cobwebs from the walls, especially in corners and cracks.

Mix 2 tablespoons of chlorine bleach into 2 gallons of boiling water. (Get an adult to help you with the water.) With the broom, scrub the inside of the coop with the bleach-water mixture. Leave the doors and windows open so the coop will dry fast. Toward evening, spread a 4-inch layer of clean litter on the floor and in the nests.

While you are at it, pick up any junk lying around outside the coop. Broken toys, scrap wood, and other trash attracts mice, rats, snakes, and insects that can harm your chickens.

Feeling a Bit Peckish

Like humans, chickens have complex nutritional needs. In the jungle, they met those needs by pecking at a variety of plants and insects. When you confine chickens to a small area, you become responsible for providing them with a balanced diet.

Chicken Feed

The best way to make sure your chickens eat right is to buy chicken feed or poultry ration at a feed store. Feed stores sell different brands, all called something different. Whatever the brand name, you can correctly call it chicken feed or poultry ration.

Feed for mature chickens is usually called *lay ration*. Lay ration contains 16 percent protein plus all the other nutrients a chicken needs. Lay ration comes either crumbled or pelleted. Crumbles have the consistency of uncooked cream of wheat. Pellets contain the same materials, pressed into the shape of a long tube, then cut into pieces resembling tiny Tootsie Rolls.

Why Use Pellets?

Chickens waste less feed if you give them pellets. Since two-thirds of the cost of keeping chickens goes into feed, the less feed you waste, the lower your cost.

Chicken Candy

Remember: Scratch is like chicken candy. If your chickens eat too much, they will get fat. Fat roosters are lazy and won't mate with your hens. Fat hens don't lay many eggs.

Grit or calcium hopper

Chicken Treats

Chickens love cracked corn or a mixture of grains called *scratch.* When you call your chickens and they come running, reward them with a handful of scratch. Use scratch as a treat while taming a chicken or training one for a show.

Throw a handful of scratch over the litter inside your hen house. Your chickens will scratch for the grain (which is how it got its name), stirring up the litter to keep it loose and dry.

Supplements

To lay eggs with hard shells that don't crack easily, a hen needs calcium in her diet. Lay ration contains enough calcium for a young hen, and hens that roam freely to find food get lots of calcium from the insects they eat.

As a hen gets older, the shells of her eggs get thin if her diet does not contain enough calcium. Feed stores sell supplemental calcium in the form of ground oyster shells or limestone. (Do *not* get *dolomitic* limestone, which can be harmful to egg production.) Place the calcium supplement in a *hopper* where your hens can eat it whenever they want it.

If your chickens free-range (see page 48), they may also need *granite grit,* which you can purchase from the feed store. Place grit in a second free-choice hopper. Since chickens have no teeth, they need grit to grind up hard things such as corn and other grains.

Included in a chicken's digestive system is a pouch called a *gizzard.* The grit a chicken eats is stored in its gizzard.

Everything a chicken eats goes through the gizzard, where it gets ground up finely enough to be digested.

Free-range chickens use pebbles and sand as grit, but they can't always find enough. A flock that does not free range and that eats only lay ration doesn't need any grit at all.

How Much Chickens Eat

One chicken eats about 2 pounds of feed each week. Dual-purpose hens eat a bit more, bantams a bit less. All chickens eat less in summer than in winter, when they need extra energy to stay warm.

Lay ration comes in 25- or 50-pound bags. To figure out how many bags you need each week, multiply the number of chickens in your flock times 2 and divide by the weight of one bag. For example, if you have twelve chickens, you need 24 pounds of feed (2 x 12 = 24). One 25-pound bag will last about 1 week. If your store carries only 50-pound bags, one bag will last you 2 weeks.

Keep a little extra lay ration on hand so you won't run out, but don't stock up too far ahead. Chicken feed goes stale, especially in warm weather. Chickens don't like stale rations any better than you like stale hamburger buns. Besides, nutritional value decreases when feed goes stale. Buy only as much feed as you will use in 2 or 3 weeks.

Safe Food

Take care not to give your chickens anything spoiled or rotten, which can make them sick. Also avoid strong-tasting foods like onions, garlic, or fish, which can make their eggs and meat taste funny.

Feed and Water Consumption

Feed per chicken	Water per chicken
2 pounds a week	2-3 cups a week

Reducing Feed Costs

Leftovers

Reduce the cost of feed by treating your chickens to leftover baked goods and vegetable scraps. Chickens love tomatoes, lettuce, apple parings, bits of toast, *cooked* potato peelings, and other tasty things from the kitchen.

Range Feeding

Another way to reduce the cost of feed is to let your chickens roam on a lawn or in a pasture for part of the day. By eating plants, seeds, and insects, they will balance their diets and eat less of the expensive commercial stuff.

Letting your chickens roam to find good things to eat is called *range feeding* or *free ranging*. When you free-range your chickens, make sure they have a source of clean water and that they can get out of the sun or rain. Also be sure they are safe from neighborhood dogs and other predators. The easiest way to free-range your flock is to leave their gate open. Then your chickens can return home when they get thirsty or tired.

Feeder Design

Chickens like to peck a little here and a little there, eating all day long. Leaving enough feed so your chickens can eat whenever they want to is called *free choice*. To feed your chickens free choice, you need a large feeder so your chickens won't run out of rations and go hungry.

The right kind of feeder keeps your chickens pecking all day. Feeders come in a variety of different designs, but good ones share certain important features. A good

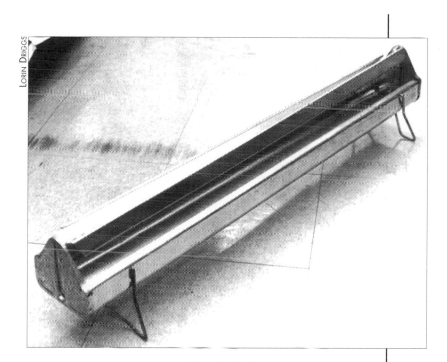

Trough feeder

design prevents chickens from using their beaks to scoop out feed, spilling it on the ground, and wasting it. Wasting feed this way is called *billing out* (even though a chicken has a beak and not a bill, like a duck).

A feeder with edges that are rolled or bent inward prevents billing out. If you have a feeder without a bent lip, prevent billing out by filling the feeder only two-thirds full. Another way to prevent billing out is to raise the feeder to the height of your chickens' backs. The easiest way to raise a feeder is to hang it from the ceiling with chains. Chains let you raise the feeder as your birds grow.

A good feeder design prevents chickens from sitting on top and messing in their feed. It is also easy to clean. Although a trough is the most common feeder design, stale or wet feed can collect in the bottom. If you use a trough, empty it and scrub it out at least once a week.

Billing out. *Using the beak to scoop feed out of the feeder onto the ground.*

The best design is a hanging feeder. It consists of a barrel-shaped container with a shallow dish at the bottom. Pour your flock's ration into the top of the barrel. As your chickens eat from the dish, feed falls into the dish from the barrel. Hang the feeder away from rain so the feed will stay clean and dry.

If you already have a feeder, or you have an idea for a new design, give it a try. Any feeder will work if it lets your chickens peck a little at a time.

How Many Feeders?

Make sure you have enough feeder space so all your chickens can eat at the same time. The lowest-ranking chickens in the peck order will go hungry if they constantly get pushed aside.

Hanging feeder

One hanging feeder is enough for up to thirty chickens. If you use a trough feeder, allow 4 inches for each bird. To figure out how long the feeder should be, multiply 4 times the number of chickens you have. For example, if you have twelve chickens, you need 48 inches of space at the feeder (4 x 12 = 48). If your chickens can eat from both sides of the trough, allow 2 inches for each bird. In this example, you would need a two-sided trough with a length of 24 inches (12 x 2 = 24).

If one feeder is not big enough for all your chickens, use two or more. As long as all your chickens have room to eat, it doesn't hurt if you have too much feeder space.

Storing Feed

Store bags of chicken feed away from moisture and off the floor or ground. The feed store may let you have a wooden or plastic pallet for storing sacks. If you can't get a pallet, place two short pieces of lumber under the sacks.

After opening a bag, store the feed in an inexpensive plastic trash container with a tight-fitting lid. A 10-gallon container holds 50 pounds of feed. A closed container keeps feed from getting stale. It also keeps out mice. Use up all the feed in the container before you open another bag. Keep the container in a cool, dry place, out of the sun.

Since mice are attracted to spilled feed, keep the floor of your storage area swept clean. One mouse can nibble a hole in your feed sack, and others will show up to eat the contents. If you suspect a mouse might be lurking about, set a mousetrap.

A 10-gallon container holds 50 pounds of feed.

Water

Water is the most important part of your flock's diet. Your chickens must be able to get a drink of water whenever they feel like it. Chickens cannot drink much at one time, so they drink often. Depending on the weather and on the size of your chickens, each bird will drink between 1 and 2 cups of water each day.

Water in hot weather. Like you, chickens get thirstier during hot weather. When the temperature gets above 80°F, your chickens may drink two to four times more than usual. Bring them fresh, cool water several times a day to make sure they get enough to drink.

Water in cold weather. In cold weather, chickens still need water, but they drink less than normal. They cannot drink frozen water and they should not be forced to eat snow. To make sure your chickens get enough to drink in cold weather, bring them warm water several times a day. If your coop has electricity, use a water-warming device to keep drinking water from freezing. Warming devices are sold at some feed stores and through poultry supply catalogs.

Waterers

Like feeders, waterers come in many different designs. The best waterer holds enough so your flock doesn't run out during the day. It keeps the water clean by not allowing chickens to step in it or to roost over it.

The worst kind of waterer is a rain puddle. Your chickens will walk in the water, mess in the water, and then drink the water. They may get sick. Even when you provide clean water, some of your chickens will drink from puddles.

Automatic watering is wonderful. With automatic watering, every time a chicken takes a sip, fresh clean water flows into the bowl. The bowl always stays full.

No Puddles Allowed

If your chicken yard has puddles in it, fill them with dirt, sand, or gravel. Since chickens dig holes while taking dust baths, one of your jobs is to fill in any hole that becomes a puddle.

Plastic 1-gallon waterer
(1) Fill the container with water. (2) Screw on the base.
(3) Flip the waterer over.

All you have to do is check every day to make sure the waterers are working, and clean them out once a week. Poultry supply catalogs carry automatic waterers. They are not expensive, but piping water to your chicken coop can be, and you may need help putting in the plumbing.

The least expensive waterer is made of plastic and holds 1 gallon. Fill the container, screw on the base, and turn the waterer over. Each time a chicken drinks, water runs out of the container through a little hole in the base.

A bit more expensive is a waterer made of metal. It holds either 3 or 5 gallons and consists of an inner container and an outer shell. The inner container has a basin at the bottom. When you put the water-filled container in place and slip on the outer shell, the shell presses against a clip that lets water flow into the basin.

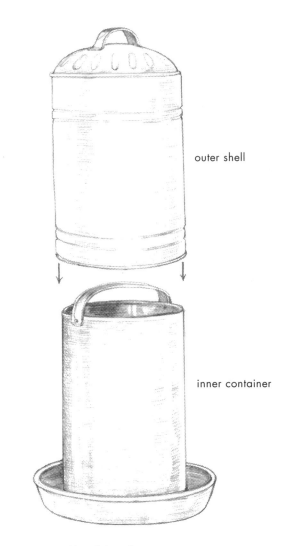

outer shell

inner container

Metal 3-gallon waterer

Making a Waterer

You can make a waterer from a 1-gallon (number 10) can. (You might be able to get this kind of can from a restaurant.) You need, in addition, a round cake pan that is 2 inches wider than the can. Punch or drill two holes opposite each other and ¾" from the open end.

Fill the can with water, cover it with the upside-down cake pan, turn the whole thing over and set it on the

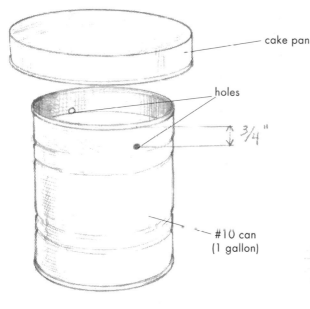

cake pan

holes

3/4"

#10 can
(1 gallon)

Home-made waterer
(1) Fill can with water. (2) Cover it with the cake pan.
(3) Turn the can upside-down.

ground. Water dribbles out of the two little holes in the can every time a chicken takes a drink from the pan.

Water at Shows

If you show your chickens, be aware that sometimes the people who run shows get too busy to remember about water. Make sure your show birds always have enough to drink by bringing along clean water from home and a small can to serve it in.

Punch a hole in the side of the can, near the open end. Thread a piece of wire through the hole and tie the can inside the show cage. (If you wish, make a second can for feed.)

When the weather is hot or you travel far, be sure your chickens have water not only during the show, but on the way there and back. And don't leave your chickens in a hot car when you arrive at the fairgrounds.

Water for a show chicken

Waterer Location

Put waterers in a shady place, so the water won't be warmed by the sun. If possible, set each waterer on top of a droppings pit. A droppings pit is so called because anything that drops on it falls into the "pit" below.

To make a droppings pit, nail together a wooden box, 42" wide by 42" long by 12" high. Staple strong wire mesh to the top of the box. Leave the bottom open. Set the box on a bed of sand or gravel, and place the waterer on top of the wire mesh.

Your chickens will stand on the wire mesh to drink. Their manure will fall through the wire, into the droppings pit. Water overflowing from the waterer will collect in the droppings pit. Your chickens will stay healthier because they can't drink dirty water from the ground.

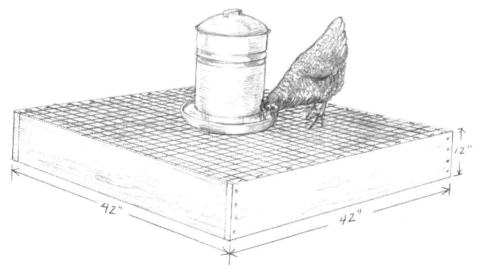

Droppings pit

The Egg and You

Eggs are a universal food, used in recipes around the world. They must be handled properly, however, to keep them fresh for eating, selling to customers, or hatching into chicks.

A Pullet's First Eggs

A pullet starts laying when she is 20 to 24 weeks old. Her first eggs are quite small, and she will lay only one egg every 3 or 4 days. By the time she is 30 weeks old, her eggs will be normal in size and she will lay about two eggs every 3 days.

A pullet starts life carrying the beginnings of as many as 4,000 eggs inside her body. Called *ova,* they are tiny, undeveloped yolks. When the pullet reaches laying age, one by one the ova grow into full-sized yolks and drop into a 2-foot-long tube called the *oviduct.*

As a yolk travels through the oviduct, it becomes surrounded by egg white and encased in a hard shell. About 24 hours after it started its journey, it is a complete egg ready to be laid.

A hen cannot lay more eggs than the total number of ova inside her body. From the day she enters this world, each female chick carries with her the beginnings of all the eggs she could possibly lay during her lifetime. However, few hens lay more than 1,000 out of the possible 4,000.

Egg Production

A good laying hen produces about 20 dozen (240) eggs in her first year. At 18 months of age, she stops laying as she molts. *Molting* is a process in which a chicken's old feathers are replaced with new ones. The feathers fall out over a period of several weeks, not all at one time. Your chicken may look ragged for a while, but rarely will one become entirely naked. After the first time, chickens molt once a year, usually in the fall. Because a hen needs all her energy to grow new feathers during the molt, she lays few eggs or none at all. Once her new feathers are in, she looks sleek and shiny, and she begins laying again.

After the first molt, a hen's eggs will be bigger than before, but there will be fewer of them. During her second year, she will lay between 16 and 18 dozen eggs. These are only averages. Some hens lay more eggs than average. Others lay fewer. Exactly how many eggs *your* hens lay depends on a number of things, including how well you manage your flock. It also depends on each individual hen. Some hens seem to work harder than others. It also depends on the breed and strain you raise. Laying breeds do better than dual-purpose breeds. Production strains lay better than exhibition strains.

How many eggs your hens lay also depends on the weather. Hens lay best when the temperature is between 45°F and 80°F. If the weather is much colder or much warmer, your hens will lay fewer eggs than usual. In warm weather, hens lay smaller eggs with thinner shells.

All hens stop laying in winter, not because the weather is cold, but because there are fewer hours of daylight in winter than in summer. The amount of daylight begins to decrease during autumn. When the number of daylight hours falls below 14, hens stop laying.

Chicken Feathers

A mature chicken has about 8,500 feathers. Growing a whole batch of new ones is quite a job. A fast molter finishes in 2 to 3 months. A slow molter may take as long as 6 months.

Keep Your Hens Laying in Winter

If your hen house is wired for electricity, you can keep your hens laying year-around by installing a 60-watt light bulb. Use the light to provide your hens with at least 14 hours of light each day, including normal sunlight. If the sun shines for 12 hours each day, for example, turn the light on for 2 hours to provide a total of 14 hours of light.

Since decreasing light discourages egg-laying, make sure the total number of light hours each day never decreases. If you forget to turn the light on, the number of light hours will decrease.

If you are afraid you might forget to turn the light on and off each day, leave it on all the time. Another possibility is to plug the light into a timer switch, which saves on electricity. Timer switches are sold by electrical supply and hardware stores. They are easy to install if your hen house has a plug-in wall outlet. A timer switch turns the light on and off at certain times each day. You have to adjust it occasionally as daylight hours change. You also have to adjust the timer whenever the power goes off for any length of time.

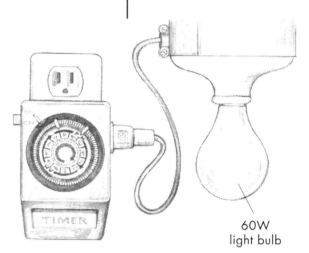

60W
light bulb

24-hour timer switch

Laying Hens and Lazy Hens

Some hens work harder at laying eggs than others. To improve the overall laying average of your flock, *cull* out the lazy layers. Culling is the process of eliminating unhealthy or inferior birds. A chicken removed from the flock because it does not measure up is called a *cull*.

When your flock reaches peak production at about 30 weeks of age, you can tell by looking at your hens and by handling them which ones are candidates for culling.

- First, look at their combs and wattles. Lazy layers have smaller combs and wattles than good layers.

- Next, pick up each hen and look at her vent — the opening under her tail where the eggs come out. A good layer has a large, moist vent. A lazy layer has a tight, dry vent.

- Now place your hand on the hen's abdomen. It should feel round and soft and pliable, not small and hard.

- With your fingers, locate the hen's two pubic bones — between her *keel* (breastbone) and her vent. In a good layer, you can easily press two or three fingers

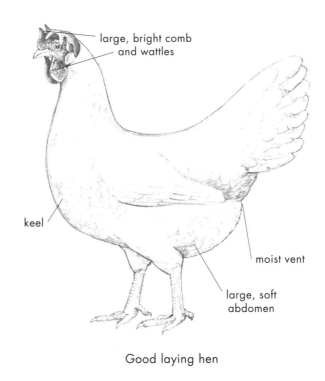

large, bright comb and wattles

keel

moist vent

large, soft abdomen

Good laying hen

Culling Check-List

Look	Layer	Non-Layer
Wattles	large	small
Comb	large & bright	small & pale
Vent	large & moist	tight & dry
Feel		
Abdomen	full & soft	shallow or hard
Pubic bones	flexible & wide apart	stiff & tight

Use this chart as a check-list for each hen. Cull any hen with more features in the non-layer column than in the layer column. By keeping only good layers, you reduce feed costs and increase laying efficiency.

between the pubic bones, and three fingers between the keel and the pubic bones. Pubic bones that are close and tight tell you the hen is not a good layer.

Skin Color and Egg Production

If you raise a yellow-skinned breed, after your hens have been laying for a while you can sort out the less-productive ones by the color of their skin. The same *pigment* that makes egg yolks yellow also colors the skin of yellow-skinned breeds.

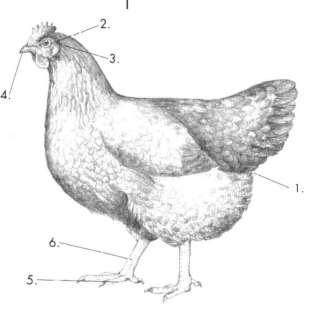

When a hen starts laying, the skin of her various body parts bleaches out in a certain order. When she stops laying, the color returns in reverse order. You can therefore tell how long a hen has been laying, or how long ago she stopped laying, by the color of the skin exposed on her beak and legs.

Bleaching sequence
1. vent
2. eye ring
3. ear lobe
4. beak (base to tip)
5. bottom of feet
6. shanks

Bleaching Sequence

Body Part	Eggs Laid	Weeks
1. Vent	0-10	1-2
2. Eye ring	8-12	2-2½
3. Ear lobe	8-10	2½ -3
4. Beak	35	5-8
5. Bottom of feet	50-60	8
6. Front of shank	90-100	10

- If you replace your hens every year, you may sell your old flock. The older hens will still lay fairly well for at least a year.

- If you replace your hens every 2 years or more, it wouldn't be fair to sell your old flock as laying hens. You could, however, sell them as stewing hens.

Replacement Pullets

A hen lays best during her first year. As she gets older, she lays fewer and fewer eggs. If you raise chickens as pets, you may not care if your hens don't lay many eggs.

If you do raise chickens for eggs, however, you have the same concern as commercial producers — a time will come when the cost of feeding your hens is greater than the value of the eggs they lay. Commercial producers rarely keep hens more than 2 years.

To keep those eggs rolling in, buy or hatch a batch of chicks every year or two. As soon as the replacement pullets start laying, get rid of the hens.

Collecting Eggs

An egg is at its best quality the moment it is laid. From then on, quality declines. Properly collecting and storing eggs slows that decline.

Pullets sometimes lay their first few eggs on the floor. A floor egg is usually soiled and may have to be "dry-cleaned" with a piece of sandpaper.

Your pullets should soon figure out what the nests are for. If one continues to lay on the floor, perhaps all the nests were occupied when she was ready to lay. Make sure you have at least one nest for every four layers. Line each nest with clean bedding to keep eggs clean.

Collect eggs often so they won't get dirty or cracked. Eggs get dirty when a hen with soiled or muddy feet enters the nest. Eggs crack when two hens try to lay in the same nest or when a hen accidentally kicks an egg as she leaves the nest.

The more often you collect eggs, the less chance they will start to spoil when the weather is hot or freeze when the weather is cold. Try to collect eggs twice a day. Most eggs are laid in the morning, so noon is a good time for your first collection. Carry them in a small bucket or a basket.

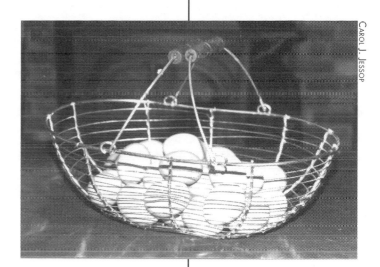

Carry eggs in a basket or small bucket.

Sorting Eggs

How you sort your eggs depends on whether you are going to eat them, sell them, or hatch them. In all three cases, throw away eggs with dirty or cracked shells. Harmful bacteria can get through the shell of a dirty or cracked egg.

If you are going to sell or hatch your eggs, sort out any that are larger or smaller than the rest. Your egg customers will be favorably impressed with cartons of eggs that are all the same size and shape. If you plan to hatch the eggs, select only eggs that are the normal size for the breed. Chicks from abnormally small or large eggs may grow up to be hens that lay eggs that are not of normal size.

Very small eggs sometimes do not contain a yolk and very large eggs sometimes have more than one. An egg with two yolks is called a *double yolker*. Occasionally an egg has three yolks or more. Such oddball eggs won't hatch and don't leave a good impression with your customers.

Whether you sell or hatch your eggs, sort out any with a weird shape or a wrinkled shell. Eggs that are too small, too large, or have an unusual shell are still good for you and your family to eat.

Eggs-tra-ordinarily Healthy Hens

According to *The Guinness Book of World Records, 1991:*

- In 1979, a White Leghorn hen at the University of Missouri laid 371 eggs in 364 days, averaging more than one egg per day.

- The world's heaviest chicken egg was also laid by a White Leghorn hen, in 1956. The egg had a double yolk and a double shell, and weighed one pound.

- The world's largest chicken egg was laid by a Black Minorca in 1896. The egg weighed nearly ¾ pound, measured 12¼ inches around the long way and 9 inches around the middle, and had five yolks.

- The record for most yolks ever found in one egg is nine. Two hens tied this record — one in the U.S. in 1971, the other in the former Soviet Union in 1977.

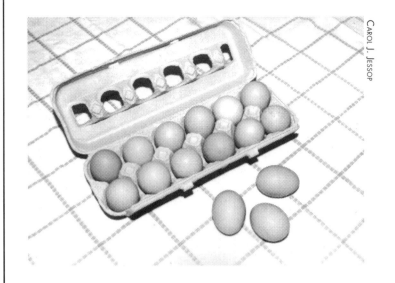

Store eggs with the larger ends up.

Storing Eggs

Store eggs in clean cartons. Place the larger end upward, so the yolk remains centered within the white.

Where you store your eggs depends on whether you are going to eat them or hatch them. If you are going to hatch them, store them in a cool, dry place but not in the refrigerator.

If you are going to sell the eggs or eat them, place them in the refrigerator as soon as possible after they are laid. An egg kept at room temperature ages more in 1 day than an egg stored in the refrigerator ages in 1 week.

The egg rack on a refrigerator door is actually *not* a good place to store eggs. Every time you open the refrigerator, eggs on the door get blasted with warm air. When you slam the door shut, you jar the contents of the eggs. Instead, keep them on the lowest shelf of the refrigerator, where the temperature is coldest. Raw eggs in a carton on the lowest shelf keep well for 4 weeks.

Egg Freshness

Fresh Eggs Don't Float

One way to tell an egg is fresh is to put it in cold water and see if it sinks. A freshly laid egg contains little air. As time goes by, moisture evaporates through the shell, creating an air space at the large end of the egg. The older the egg, the larger the air space. If the air space is big enough to make the egg float, the egg is too old to eat.

old egg floats

fresh egg sinks

How Fresh Eggs Look

When you break a fresh egg into a dish, the white is compact and firmly holds the yolk up. In an old egg, the white is runny and the yolk flattens out.

Compare one of your hen's eggs with a store-bought egg. Which egg is freshest?

old yolk flattens

fresh yolk stands up

Why Eggs Spoil

Storage Condition	Result
High temperature	White breaks down; any bacteria present multiply.
Low humidity	Evaporation enlarges air cell.
High humidity	Bacteria and molds grow.
Dirty shell	Carries bacteria.
Rough handling	Shell cracks, contents spoil.

Candling Eggs

Yet another way to tell an egg is fresh is to use a light to examine the yolk and white and the air space at its large end. This was once done with a candle. These days an electric light is used, but the process is still called *candling* and the device used is called a *candler*.

You can buy an egg candler from a poultry supply catalog or you can make one. To make a candler, all you need is a strong flashlight and a sturdy cardboard box with a hole in it, about the size of a quarter. The illustration shows how to put them together. Make sure no light comes out of the candler except through the hole.

Use your candler in a dark room. Grasp each egg by its small end and hold it at a slant, with the large end against the hole.

When you examine an egg with a candler, you can see its contents through the shell. In a fresh egg, the air space is no more than ⅛" deep. The yolk is a barely visible shadow that hardly moves when you give the egg a quick twist.

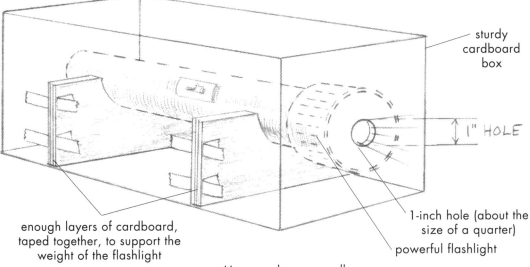

sturdy cardboard box

1" HOLE

enough layers of cardboard, taped together, to support the weight of the flashlight

1-inch hole (about the size of a quarter)

powerful flashlight

Homemade egg candler

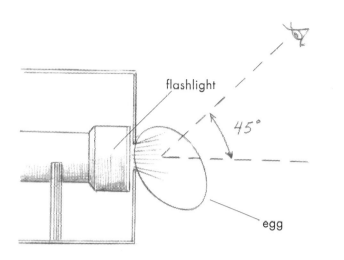

flashlight

45°

egg

Using a homemade candler

In an old or stale egg, the air space is large and sometimes irregular in shape. The yolk is a plainly visible shadow that moves freely when you give the egg a twist.

Occasionally, you may see a small, dark spot near the yolk or floating in the white. It is a bit of blood or flesh that got into the egg while it was being formed. Blood spots and meat spots are harmless, but they don't look nice. When you sort eggs for sale, eliminate any with spots.

If you aren't sure what you are looking at when you candle an egg, break the egg into a dish and examine it. Soon you will become confident enough to candle eggs like a pro.

Nutritional Value of Eggs

Eggs have been called the perfect food. One egg contains almost all the nutrients necessary for life. The only essential nutrient it lacks is vitamin C.

The average egg weighs about 2 ounces. The shell accounts for approximately 10 percent of the weight, the egg white 60 percent, and the yolk 30 percent.

Cholesterol

Most of the fat, as well as all of an egg's cholesterol, is in its yolk. To reduce the cholesterol in an egg recipe such as scrambled eggs or omelettes, use 2 egg whites instead of 1 whole egg for half the eggs in the recipe. For example, if the recipe calls for 4 eggs, use 2 whole eggs plus 4 egg whites.

To eliminate cholesterol in a recipe for cakes, cookies, or muffins, substitute 2 egg whites and 1 teaspoon of vegetable oil for each whole egg in the recipe. For example, in a recipe calling for 2 eggs, use 2 egg whites plus 2 teaspoons of vegetable oil. If the recipe already has oil in it, leave out the extra 2 teaspoons.

Fertile Eggs

Many people believe that if you keep a rooster with your hens, the fertile eggs your hens lay are more nutritious than infertile eggs. They aren't.

The eggs sold in most supermarkets are not fertile. Some sellers try to fool their customers into paying more by advertising fertile eggs as being more nutritious. If you sell your eggs, you can get a few cents more than the store charges, not because your eggs are fertile, but because they are homegrown and taste better than store-bought eggs.

Egg Color

Each hen's eggs have a specific shell color. All her eggs might be white, brown, speckled, blue, or green. The color of the shell has nothing to do with the nutritional value of the egg.

White-shelled eggs. Hens of the Mediterranean breeds lay white-shelled eggs. Since Mediterraneans are the most efficient layers, they are preferred by commercial egg producers. Some customers like eggs with white shells because that is what they are used to.

Brown-shelled eggs. Brown-shelled eggs are laid by American breeds. Since the Americans are dual-purpose, they are popular in backyard flocks. Some customers prefer brown eggs because they look home-grown. Brown eggs come in every shade from dark reddish to light tan. Some tan shells are so pale they look pink. Some brown eggs have dark speckles on them, like some people have freckles on their noses. You know an egg is speckled, rather than dirty, if you can't rub the spots off.

Blue and green eggs. Blue-shelled eggs are laid by a South American breed called Araucana and its relative, the Ameraucana. These breeds lay such pretty eggs they are sometimes called Easter-egg chickens. Green-shelled eggs are laid by hens bred from a cross between Araucana and a brown-egg breed. Customers who enjoy unusual things get a kick out of eggs with blue or green shells. Sellers sometimes charge outrageous prices for blue or green eggs, claiming they are lower in cholesterol than white or brown eggs. *You* know that can't be true, since the color of the shell has nothing to do with the nutritional value of the egg.

Cooking Eggs

All the nutrients that make an egg good to eat also create a good environment for harmful bacteria to grow, so it is important to cook only clean eggs. Compared to dirty eggs, clean eggs start out with fewer bacteria on their shells. Prompt refrigeration is important because cold temperatures slow the growth of bacteria. Any bacteria present can be destroyed by thorough cooking.

One safe way to cook an egg is to hard-boil it. Any time you cook an egg some other way, make sure the white is cooked through and the yolk begins to thicken. The yolk need not be cooked hard, but it should not be runny.

Cook all eggs slowly to make sure they are heated all the way through. Eat an egg as soon it is cooked.

Microwaving Eggs

You can use a microwave oven to fry or scramble eggs. For fried eggs, break 2 eggs into a greased microwave-proof pie plate. Prick each yolk with a toothpick or the tip of a sharp knife. If you do not prick the yolks, steam will build up inside the yolks and they will explode. What a mess!

Cover the plate with plastic wrap. Place the plate in the microwave on 50 percent power for 2 to 3 minutes, or until the eggs are almost cooked. Leave the plate in the microwave for another 30 to 60 seconds, until the eggs are completely cooked.

For scrambled eggs, break 2 eggs into a large microwave-proof cup. Beat them with 2 tablespoons of water and a dash of salt. Place the cup in the microwave on full power for 60 to 90 seconds, until the eggs are almost done, stirring once. Cover the cup with plastic wrap and leave it in the microwave for 1 minute more, until the eggs get firm.

Hard-Cooked Eggs

Hard-cooked eggs are often called "hard-boiled," even though they shouldn't really be boiled. The one time you don't want to use fresh eggs is when you hard-cook them. Since the contents of a fresh egg fill up the shell, the shell sticks and is difficult to peel off when you hard-cook it. After some of the moisture evaporates from an egg, the contents shrink away from the shell. If the egg is at least 1 week old, the shell will peel off easily.

Another trick is to avoid cracking the shell. A cold egg cracks when you drop it into boiling water. To avoid this, place cold eggs in a saucepan, cover them with cold water, then bring the water to a boil. Turn off the stove, cover the pan, and leave the eggs in the hot water for 10 minutes.

A third trick is to keep the yolk from turning green around the edges. A green yolk isn't harmful, but it doesn't look appetizing. As soon as the eggs are done, cool them quickly and the yolks won't turn green. Either put the eggs in a pan of ice water or run cold faucet water over them.

An unpeeled hard-cooked egg looks just like a raw egg. So you won't confuse them in the refrigerator, use a pencil to mark boiled eggs with an X. If you forget to mark your eggs, you can tell if one is cooked or raw by spinning it like a top. The contents of a cooked egg are solid, while the contents of a raw egg are liquid. A hard-cooked egg is therefore easier to spin.

The Artful Egg

People have been decorating eggs since ancient times. Egg decorators are called *eggers*. You can learn all about egg art and find out about exhibits (*eggs*ibits!) in your area from the *Egger's Journal* or The National Egg Art Guild. Another good source is the book, *Easter Eggs — By the Dozens!* (Sources for these are at the back of this book.)

Keep Dyed Eggs Safe

When you color boiled eggs as a holiday decoration and leave them sitting around at room temperature, they are no longer safe to eat. In the refrigerator, cooked eggs keep well for 1 week.

Egg art

Pink
Beet parings,
cranberries

Blue
Red cabbage leaves, red
onion skins

Greenish gold
Yellow Delicious apple
parings

Pale green
Spinach leaves

Yellow
Orange rind,
lemon rind, carrot tops

Orange
Yellow onion skins

Brown
Coffee grounds,
tea leaves

Eggs are decorated after being either hard-cooked or blown. They can be dyed, painted, colored with crayons or felt-tipped pens, turned into funny faces, topped with hats, trimmed with feathers, or decorated zillion other ways. Hard-boiled eggs are sturdier for handling. Blown eggs last longer and make good decorations.

Blowing a Raw Egg

To blow a raw egg, first wash and dry the shell. With a pin, carefully prick a hole in both ends. Insert a long needle through one hole and twist it around until the contents are scrambled.

Hold the egg over a sink and blow into one hole, forcing the contents out the other hole. Emptying a shell takes time, so be patient.

Dyeing Eggs

Dyeing is the easiest way to decorate an egg. White eggs are easier to dye than colored eggs. Egg-dyeing kits are available at grocery or department stores around Easter time, but making your own natural egg dyes is more fun.

Measure out enough water to cover the eggs in a saucepan. Add 1 tablespoon of white vinegar for each cup of water. Then add vegetable scraps from your kitchen or garden. The more scraps you add, the darker your eggs will get.

Simmer the eggs for 20 minutes. If you are dyeing blown eggs, the shells will float, so keep turning them to get an even color.

Which Came First?

No one will ever solve the age-old riddle, "Which came first, the chicken or the egg?" We do know that a hen lays eggs to make more chickens.

Eggs for Hatching

An egg contains everything a chick needs for survival while it is developing inside the shell. Actually, a chick inside the shell is more properly called an *embryo*. An unborn human baby is also called an embryo. But unlike a human baby developing inside its mother's body, a chick develops outside its mother's body. Because it comes out of an egg, a chick *hatches* instead of being born.

An egg must be fertile to hatch. If you keep a rooster, chances are good that your hens will lay fertile eggs. Eggs get fertilized when a rooster's *sperm* joins with one ovum in a hen's body. Under the right conditions, a joined sperm and ovum eventually grow to become a chick. How well you manage your breeding flock determines how the chick turns out.

Your Breeding Flock

A breeding flock is a group of chickens from which eggs are collected for hatching. The flock must include at least one cock so the eggs will be fertile.

You don't *have* to keep a breeding flock. If you raise chickens for meat, you might buy chicks each spring and keep them for only 2 or 3 months, until they are big enough to eat. Even if you keep chickens for eggs, when your hens stop laying well, you can simply buy more chicks to raise instead of hatching them.

There are good reasons for keeping a breeding flock, however:

- If you raise endangered chickens, with a breeding flock you can try to get your breed off the endangered list.

- You can enjoy the challenge of improving your chickens. If you keep chickens for eggs, you might breed hens that lay more eggs. If you keep chickens for show, you might improve your chance of winning a prize.

- By far the best reason to keep a breeding flock is because hatching eggs from your own chickens is fun.

Selecting Mates for Breeding

If you want consistent results in your chicks, your breeding flock must contain purebred chickens of one single breed. "Consistent results" means all the chicks grow up to be like the chickens in your flock. "Purebred" refers to any chicken whose mother and father are both the same breed and variety.

You won't get consistent results if you hatch eggs from commercially crossbred chickens. To get chicks like your chickens, you must use the same mating pattern as the person you bought your original chickens from. That pattern may be quite complex, making this a good project to save until you gain more experience.

You won't get consistent results if you hatch eggs from barnyard chickens or from a flock containing more than one breed. People sometimes hatch eggs from mixed breeds to see what the chicks will look like, but that is not a very challenging goal. Besides, if you want to sell chicks, you won't get as much for mixed breeds as for purebreds.

In addition to being purebred, the birds in your breeding flock should be healthy and free from deformities. They should be true to type, meaning each bird is of the correct size, shape, and color for its breed. Cull any chicken from your breeding flock that doesn't measure up.

Buying Hatching Eggs

Instead of breeding your own chickens, you can buy *hatching eggs*. Buying hatching eggs is great fun, but it is also risky business. The eggs may be too old to hatch. They may have been stored or handled improperly. They may not be fertile. Few sellers guarantee that their eggs will hatch. When you buy hatching eggs, you pay your money and you take your chances.

The Best Time to Hatch

The best months to hatch are February and March, unless you live in an extremely cold climate. Then March and April are the best months. A breeding flock is strongest and healthiest in spring, so its chicks are also strongest and healthiest in spring.

Pullets hatched in spring will lay by fall and will continue laying for a year. Pullets hatched in winter will lay by midsummer, but may molt and stop laying in the fall. Chicks hatched in summer are not yet strong enough to ward off disease-causing organisms that flourish in warm weather.

Incubation

You can hatch eggs in two different ways. Let a hen hatch them for you or hatch them in a mechanical device. Either way, the process is called *incubation*.

If you let a hen handle the job, it is called *natural incubation*. The hen doing the hatching is called a *setting hen* or a *broody hen*. The eggs in the hen's nest are called a *setting* or a *clutch*. The chicks that hatch are also sometimes called a *clutch*, but more often they are called the hen's *brood*.

Incubation. The process of hatching fertile eggs.

If you hatch eggs in a mechanical device, the process is called *mechanical* or *artificial incubation*. The device in which the eggs are hatched is called an *incubator*. An incubator is designed to imitate the temperature and humidity produced by a setting hen.

Incubators

Incubators come in several styles and a wide range of prices. They are sold through feed stores and poultry supply catalogs.

Some incubators have a window in the cover so you can watch your eggs hatch. Some have a fan that circulates warm air so all the eggs stay the same temperature.

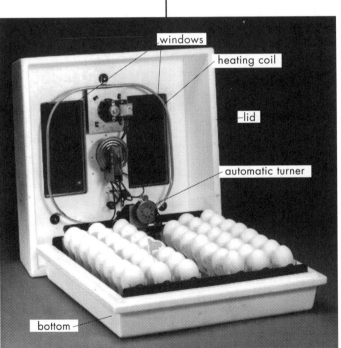

MILLER MANUFACTURING

windows

heating coil

lid

automatic turner

bottom

Incubator with windows, a fan, and automatic turner

Some incubators have an automatic turner. An incubator without an automatic turner costs less, but you have to turn the eggs several times a day by hand. It is easy to get busy with other things and forget to turn your eggs. Turning keeps the embryo floating in the egg white so it won't stick to the inside of the shell. When a hen hatches eggs, she constantly fidgets in the nest, causing her eggs to turn.

If you get an incubator with a window *and* automatic turning *and* a fan, you will really have it made. Every incubator comes with a set of directions. Follow them carefully. Your county Extension Service agent may also have helpful information.

Hatching Egg Selection

For hatching, select only eggs that are clean and the proper size, shape, and color for your breed. Eliminate eggs that are cracked, unusually large, unusually small, round, or oblong.

Keep It Closed

Don't open your incubator unnecessarily. It will cause humidity to escape, which causes the eggs to dry out and not hatch as well.

Manual Turning Record

Date	Time Eggs Turned	Room Temp.	Incub. Temp.	Remarks

Use your egg candler to check the eggs for blood spots. Blood spots can be hereditary, meaning the daughters of a hen that lays eggs with blood spots may also lay eggs with blood spots.

Depending on how many eggs your incubator holds and on how well your hens lay, it may take several days to gather enough eggs to fill your incubator. You can store hatching eggs for up to ten days. Store them in a clean egg carton with their large ends upward. Keep the eggs out of sunlight, in a cool place but *not in the refrigerator*. The best storage temperature is about 55°F. Use a thermometer to find a room in your house with the right temperature. If your house has a basement or a pantry, check there first.

A hen also has to collect a *clutch* of eggs before she starts hatching them. If she incubated each egg as she laid it, the first chick would be ready to leave the nest before the last chick was ready to hatch.

Clutch. *A batch of eggs that are hatched together.*

Hatching Record

Date Set	Number of Eggs Set	Number of Eggs Fertile	Percentage of Eggs Fertile	Number of Eggs Hatched	Percentage of Eggs Hatched	Date Hatched

Hatching Rate

No matter how careful you are, not every egg will hatch. Some eggs won't hatch because they are infertile. Maybe your rooster has favorite hens and didn't get around to mating with the rest. Maybe he is too young, or too old. Maybe he is ill. If you have more than one rooster, maybe they spend too much time fighting.

Supposing the eggs *are* fertile, even a hen may not get them all to hatch. A good hen, however, hatches a higher percentage of eggs than you can hatch in an incubator.

The average hatching rate in a good incubator is about 85 percent of all fertile eggs. If your incubator holds 60 eggs, 85 percent is 51 eggs. If you hatch 51 chicks or more from 60 eggs, you are doing well.

Natural Incubation

Wild jungle fowl hens hatched their eggs without any help. Then people came along. Instead of leaving eggs in a hen's nest, they took the eggs away. The hen laid more eggs, trying to get enough together to hatch. But the people didn't want their hens to set. When a hen starts setting, she stops laying.

Over the years, people who kept laying hens hatched eggs only from hens that were more interested in laying and less interested in setting. The daughters of those hens were even less interested in setting than their mothers. Today, the best laying breeds are the least likely to brood. Hens deposit their eggs in a nest and then leave. In commercial egg production, hens live in cages and never even see a nest.

Luckily, the hens of some breeds are still good setters. The breeds most likely to brood are Cochin, Orpington, Old English Game, Plymouth Rock, Rhode Island Red, Wyandotte, and Silkie. Silkie hens do such a good job that they are often used to hatch the eggs of rare and exotic birds.

Why Eggs Don't Hatch

- Breeding flock unhealthy, weak, or improperly fed
- Infertile eggs
- Dirty eggs
- Eggs improperly stored
- Eggs too old
- Eggs not turned often enough
- Incubation temperature too high, too low, or not steady
- Incubation humidity too low or too high
- Incubator not properly ventilated

Silkies are good broody hens.

Managing a Broody Hen

When a hen is beginning to set, she stays in the nest long after she has laid her egg. If you touch her or try to take her eggs away, she puffs up her feathers and makes growling sounds. She might peck your hand. Within 2 or 3 days she will begin to brood for real. She will stay on the nest for 21 days, until her eggs hatch.

A setting hen gets off the nest for only a few minutes a day to grab a bite to eat. While she is out to lunch, another hen may come along and lay an egg in her nest. That's not a problem for the broody hen. She will hatch that egg, too.

However, if the broody hen comes back while the other hen is still in her nest, she may get confused and go into a different nest. The other hen lays her egg and leaves. The partially incubated eggs get cold. The embryos die.

Make sure your broody hen does not have lice or mites when she starts to set. Check her often and see that she remains lice- and mite-free. (See Chapter 8 for details.)

When one of your hens gets broody, separate her from your other chickens. Move her to a private place where she can brood without being bothered.

Move the broody and her eggs at night. If you move her during the day, she may try to get back to her old nest. When you move her at night, she will wake up in the morning and find herself in a dark, quiet place. Chances are good she will stay there.

Of course, chickens are like people — you can't always predict what they will do. After you move a hen, she may stop setting. Even if you *don't* move her, she may stop setting before her eggs hatch. The best you can do is make sure your broody hen has a comfortable, quiet place and enough to eat and drink, and then let nature take its course.

Egg Anatomy

An egg consists of many parts. First there is the shell. If you look closely, you can see tiny *pores,* which allow oxygen and carbon dioxide to pass through the shell. More pores are at the big end than at the little end, so there is more air at the big end than at the little end.

The shell is lined with a leathery outer *membrane.* Within the outer membrane is an inner membrane. These membranes help keep bacteria from getting into the egg and they slow the evaporation of moisture from the egg.

Between the outer membrane and the inner membrane, at the larger end of the egg called the *air-cell end,* is an air cell. The air cell holds oxygen for the chick to breathe.

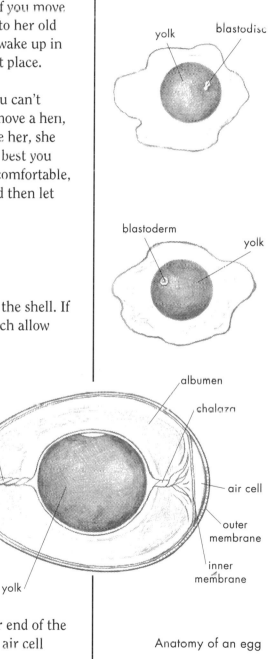

Anatomy of an egg

The inner membrane surrounds the egg white, called *albumen*. Albumen is 88 percent water and 11 percent protein. One of the functions of the albumen is to cushion the egg yolk that is floating in it. The yolk is made up of fats, carbohydrates, proteins, vitamins, and minerals to feed the growing embryo.

On two sides of the yolk are cords that keep the yolk floating within the albumen. These cords are called *chalazae* (one cord is a *chalaza)*. When you crack an egg open, the chalazae break, recoiling against the yolk. You can easily see them. They look like white lumps on two sides of the yolk. Many people mistakenly think these white lumps are the beginnings of a baby chick. Now *you* know what they really are.

On top of the yolk is a round, whitish spot, called the *germinal disc* or *blastodisc*. If the egg is infertile, the blastodisc is irregular in shape.

When an egg is fertile, the blastodisc is called a *blastoderm*. A blastoderm looks like a set of tiny rings, one inside the other. During incubation, the blastoderm develops into a baby chick.

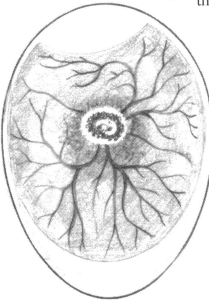

Embryo seen through the shell of a partially incubated egg during candling

Egg Fertility

If you break an egg into a dish and look very carefully at the yolk, you can tell whether or not the egg is fertile. An infertile egg has an irregularly shaped blastodisc. A fertile egg has a perfectly round blastoderm. You may need a magnifying glass to see the difference.

After the egg has been broken, of course, it can no longer be incubated. Unfortunately, there is no way to tell from the egg's outside whether or not it is fertile.

You can, however, tell if an egg is fertile after it has been incubated for one week. Hold the egg against your candler and gently turn it to see what is inside. If the egg looks clear, it is infertile. If the egg is fertile, you will see a dark spot at the center — a newly formed heart! Look carefully and you will see the heart beating.

Surrounding the heart are blood vessels that carry food and water from the yolk to the developing embryo.

Embryo Development

A blastoderm needs food, water, and oxygen in order to develop into an embryo. It gets food and water from the egg yolk. It gets oxygen from air coming through the pores in the shell.

Week 1. Within the shell, the embryo has everything it needs to develop into a chick. By the second day, it has a heart. By the third day, it has a head with eyes and little wings and legs. By the end of the first week, the embryo has all the body parts of a finished chicken, although they are still undeveloped. It even has the beginnings of feathers.

Week 2. By the ninth day, the embryo begins to look like a chick. By the fourteenth day it even has feathers.

Week 3. At the beginning of the third week, the embryo turns its head toward the air-cell end of the egg. It continues to grow throughout the week, absorbing the remainder of the yolk.

By the twentieth day, the embryo occupies the entire space inside the shell, except the air cell. It needs more oxygen, so it uses its beak to break through the inner membrane into the air cell. There, it takes its first real breath and is so happy it peeps. You can hear it peeping inside the shell.

On the twenty-first day, the chick makes a tiny hole, or *pip,* in the shell. It does this using its *egg tooth,* a sharp projection at the top of its upper beak. The egg tooth has only one purpose. Soon after the chick hatches, the egg tooth falls off.

After using its egg tooth to pip the shell, the chick rests for 3 to 8 hours. Then it turns its head and breaks the shell all the way around. As it chips away at the air-cell end of the shell, the chick shoves against the small end with its feet. Finally, after about 40 minutes of hard work, the chick kicks free and lies there, wet and exhausted.

Hatching Data

Best temperature to store hatching eggs:
55°F

Incubation period:
21 days

Best hatching time:
February-March
in the South

March-April
in the North

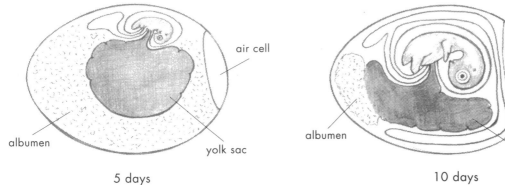

5 days

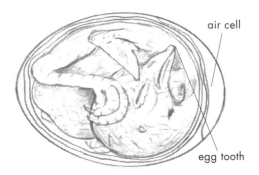

10 days

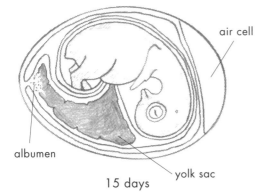

15 days

20 days

21 days

Embryo development

From Egg to Chick

When you hatch chicks in an incubator, you are responsible for providing them with warmth, protection, and food. If your hen hatches her own chicks, your job is to care for the hen so the hen can properly care for her chicks.

The Mother Hen

When a hen hatches her own brood, she keeps them in the nest until she is sure they are ready to venture out into the world. Even after they leave the nest, she keeps them warm and helps them find food. A mother hen gathers her brood under her wings if it rains or if she thinks her chicks are in danger. She squawks and puffs herself up as big as possible to chase away any dog, cat, or human that comes near.

   If you hatch chicks in an incubator, you become the mother hen.

Chick Care

Leave newly hatched chicks in the incubator until they are completely dry and fluffy. By that time, they will be scrambling around and complaining loudly. They are telling you they are ready to come out into the world.

When you open the incubator, you may find that some chicks have hatched late and are still wet and tired. Leave them in the incubator a few more hours. Do not remove wet chicks or they may chill.

Keep your chicks in a warm, dry, draft-free place where they are protected from dogs, cats, and other animals. The place where you keep your brood of chicks is called a *brooder*. The simplest brooder is a large, sturdy cardboard box. Find or make a box that is big enough to provide 6 square inches of space for each chick.

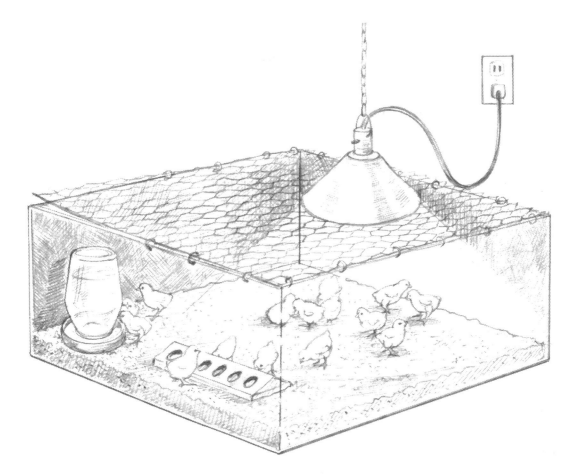

Brooder box

Fasten a piece of chicken wire to the top of the box so your chicks can get air but pets can't get your chicks. At one end of the box, place a light bulb in a reflector. A reflector with a screw-in light socket costs just a few dollars at the hardware store.

Place an inch of peat moss, shavings, sand, or other litter at the bottom of the brooder. Litter keeps chicks warm and dry and absorbs their droppings. It gives chicks something soft to sleep on and a rough surface to walk on. Each day, sprinkle a little clean litter over the old litter.

For the first two days, cover the litter with paper towels so your chicks won't try to eat it. Don't use newspaper, or your chicks will slip around and have a hard time walking.

Chick Comfort

Heat from the light will keep your chickens warm. You can tell they are comfortable by the way they act. If they are warm enough, they wander freely around the box. When they sleep, they spread out like a carpet.

Chicks that are not warm enough crowd under the light and cheep loudly. They sleep in a pile and may smother each other. To warm your chicks, make sure the brooder is not drafty and try a stronger light bulb.

When chicks are too warm, they move as far away from the light as they can. They may crowd into the corners of their brooder, possibly smothering one another. Reduce the wattage of the light bulb, raise the reflector, or get a bigger box where your chicks can get farther away from the heat.

Feeding Chicks

A chick should drink its first water soon after it comes out of the incubator, but it may not eat right away. It is still living on reserves supplied by the yolk it absorbed just before hatching.

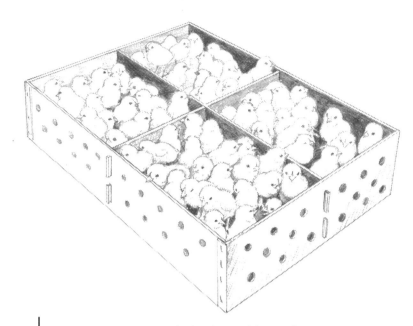

Chicks shipped by mail

When you order chicks by mail, they are shipped as soon as they are ready to leave the incubator. Their yolk reserves let them survive many hours in the mail. By the time they arrive, though, they will be ready for food and water.

Feed your chicks a *starter ration* purchased from the feed store. Starter is higher in protein and lower in calcium than lay ration. Some brands of starter are medicated, some are not. If you take proper care of your chicks and keep their housing clean, medication is not necessary. Medicated feed is a poor substitute for good management.

In areas where chickens are big business, feed stores sell a variety of rations for chicks. You might find starter ration for newly hatched chicks and *grower ration* for older chicks. You might find one kind of grower ration for meat birds and another kind for layers. In most parts of the country, though, you will find one all-purpose starter or starter-grower. When your chicks are 18 weeks old, gradually mix a little lay

No Lay Ration

Never feed lay ration to young chickens. The extra calcium it contains can seriously injure their kidneys.

ration into their starter. Completely switch them over to adult rations by the time all your pullets are laying, at about 22 weeks of age.

Chick Feeders

Place the starter ration in a feeder designed especially for chicks, and make sure food is always available. Find chick feeders at a feed store or through a poultry supply catalog.

If the feeder has a cover with slots in it, allow one slot per chick. If the feeder does not have a cover, allow 1 inch per chick. If your chicks can eat from either side of the feeder, figure both sides into your calculations.

To minimize waste, fill the feeder only two-thirds full. Keep the feeder the height of the chicks' backs, raising it as the chicks grow. One way to raise the feeder is to hang it from wires or chains. Another way is to fasten a block of wood to the bottom.

As they grow, your chicks will become too big to eat from a chick feeder. To make sure they don't go hungry, be ready to switch to a chicken-sized feeder.

Chick feeders

Feeding Broiler Chicks

When you raise chicks for eggs, you want them to grow slowly so they are fully developed by the time they start to lay. When you raise broilers, you want them to grow as fast as possible so they are still tender by the time they get big enough to eat. The younger the chicken, the more tender it is.

Young birds convert feed into meat more efficiently than older ones. The most economical meat is a broiler or fryer weighing between 2½ and 3½ pounds. If you want a bigger bird weighing 4 to 6 pounds to roast like a turkey, it will cost you more per pound to raise.

Feed your roaster or broiler chicks often to stimulate their appetites and encourage them to eat more. The more they eat, the faster they grow.

If you feed medicated rations to your broilers, read the label to learn the *withdrawal period*. It is likely to be 30 days. It takes that long after you stop giving medicated feed to your broilers for the medication to disappear from their meat. To make sure you don't eat medicine along with the meat, you must find a source of non-medicated ration to feed your broilers during the withdrawal time.

How Much a Chick Eats

Broilers

An efficient broiler eats approximately 2 pounds of starter for every pound of weight it gains. If you raise your broilers to 3½ pounds, each one eats 7 pounds of starter. To figure out how much total starter you need, multiply the number of broilers you have times 7. If you raise 25 broilers you need a total of 175 pounds of starter ($7 \times 25 = 175$).

If you raise an efficient meat breed or hybrid, your broilers will be ready to eat in 7 to 8 weeks. If you raise a dual-purpose breed, your broilers won't grow as rapidly. Depending on their breed, your birds could eat twice as much as a meat breed by the time they weigh 3½ pounds.

Layers

If you raise one of the laying breeds, each pullet will eat about 25 pounds of feed before she starts laying at about 20 weeks of age. Suppose you have 25 pullets: 25 x 25 = 625 pounds of feed, the amount your pullets will eat before laying their first eggs.

The dual-purpose breeds take longer to reach laying age. During that extra time, each pullet eats about 2 pounds of extra feed. By the time 25 dual-purpose hens start to lay, they eat 50 pounds (2 x 25 = 50) of feed more than hens of an efficient laying breed.

If the dual-purpose breeds eat more than the egg or meat breeds, why would anyone raise them? Well, if you start with a batch of straight-run dual-purpose chicks, you can raise the cockerels as broilers and the pullets as layers. It is a little like getting two for the price of one.

Water

Chicks need water at all times. The water must be clean and fresh if you want your chicks to be healthy.

The easiest way to water your chicks is to use a 1-quart mayonnaise or canning jar. At the feed store, or through a catalog, you will find a chick watering basin that fits the top of the jar.

Fill the jar with water, place the basin on top, and flip the jar over. Every time a chick takes a drink, water flows out of the jar into the basin. Your chicks won't be able to walk in the water and get it dirty or fall in it and drown.

Chick waterer

As your chicks get older, a 1-quart jar becomes too small to provide them with enough water, and the little basin becomes harder for them to drink from. Switch to a 1-gallon waterer purchased from a feed store or poultry supply catalog, or make one yourself following the directions in Chapter 4.

Health Problems of Chicks

Different chick diseases occur in different parts of the country. Your county Extension Service agent or state poultry specialist can tell you whether or not your chicks should be vaccinated.

Pasting Up

A fairly common problem in newly hatched chicks is *pasting up*. Pasting up occurs when chick droppings stick to the bird's rear end. After a while, the vent gets pasted shut. Gently pick off the wad of hardened droppings, taking care not to tear the chick's tender skin.

If several chicks paste up, it may be because they have gotten chilled. Take steps to make the brooder warmer.

If pasting up continues even though your chicks are warm enough, mix a little cornmeal or ground-up raw oatmeal with their starter. By the time your chicks are 1 week old, pasting up should no longer be a problem.

Coccidiosis

Another common problem to watch out for is *coccidiosis* (pronounced COCK-sid-e-O-sis). When chicks get coccidiosis, their droppings turn loose and watery.

Sometimes blood appears in the droppings.

Coccidiosis occurs more often during warm humid weather. Chicks raised in the cool weather of early spring are unlikely to get it unless they live in dirty conditions or are forced to drink dirty water.

To prevent coccidiosis, wash your chicks' water basin every time you fill the jar. If you find chick droppings in the water basin, dump it out and refill it with fresh water.

Keep the brooder box lined with clean litter. When the litter gets dirty or wet, change the litter or start over with a fresh box.

If you plan to raise chicks regularly, avoid coccidiosis by using a permanent brooder with a raised wire floor. Droppings fall below the wire, where the chicks can't peck in them. Buy a metal brooder from a feed store or poultry supplier, or make a wooden one.

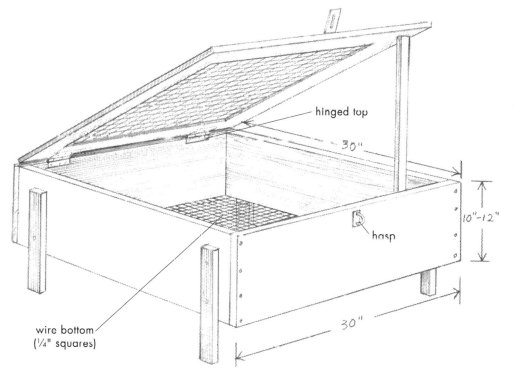

Wooden brooder

Another way to keep your chicks from getting coccidiosis is to feed them medicated starter. Medicated starter contains a *coccidiostat* — a drug that prevents coccidiosis. Medicated feed won't help if your chicks already have coccidiosis. In that case you need stronger medication, purchased from a feed store or poultry supplier.

If you keep your chicks warm, dry, and away from drafts and predators, feed them properly, and make sure they always have clean water, chances are good they will thrive as they grow.

Special Care for Meat Breeds

Don't use roosts if you raise broilers. A roost causes heavy birds to get blisters (enlarged bubbles) on their breasts. Blisters and callouses (thickened skin) also occur on heavy meat birds housed on a wire floor or on packed, damp litter.

From Chicks to Chickens

As your chicks grow, they need less heat and more space. If you keep them too warm or crowded, they may peck at each other and cause serious wounds.

Your chicks will start growing feathers on their wings within a day or two of hatching. Depending on their breed, they will be fully-feathered by the time they are 4 to 6 weeks old. By then they need at least 1 square foot of space each, and they no longer need artificial heat. They are ready to be moved out of the brooder and into the chicken house.

Learning to Perch

If they have a place to perch, your chicks will practice perching when they are only a few days old. At first they perch for only a few minutes at a time. One or two chicks may play the perching game — a chick jumps onto the perch and jumps back off the other side, scaring the dickens out of all the other chicks.

After a few weeks, some of your chicks will roost on the perch overnight. By the time they are 4 or 5 weeks old, all the chicks will roost. Allow 4 inches of roosting space per chick.

Culling

As your chicks grow, cull any that are deformed. Watch especially for runts and chicks with crooked breasts or backs. Deformed chickens do not grow well, may not lay well, cannot be shown, and should not be kept for breeding.

Physical Maturity

Your chicks will develop reddened combs and wattles when they are 3 to 8 weeks old. Some breeds develop earlier than others. Cockerels have larger, more brightly colored combs and wattles than pullets.

It won't be long before your cockerels try to crow. At first they will sound pretty funny. Don't worry. With a little practice, they will get the hang of it.

Chicken Development

Chicks are fully feathered:	4-6 weeks
Cockerels learn to crow:	6-8 weeks
Pullets start laying:	20-24 weeks

When your cockerels start chasing your pullets, it is time to separate them. Select your best cockerels for breeding and raise the rest as broilers.

Your pullets should start laying when they are about 20 weeks old. At first their eggs will be quite small and will likely be laid on the floor. After a few weeks, your pullets will find the nests and will start laying regular-sized eggs.

Chick Management Schedule

Before chicks hatch: Organize brooder, heat, litter, feeder, waterer, starter ration.

After chicks hatch: Place chicks in brooder.
Watch to see that they aren't too hot or too cold.
Watch to see that all know how to eat and drink.

Daily: Clean and refill waterer.
Keep feeder filled with starter.
Remove wet litter, add fresh litter.
Check for pasting up.
Lower temperature slightly if chicks appear too warm.

Third week: Increase brooding space.
Supply additional source of water.
Supply additional source of feed.

Fifth week: Turn off heat if weather is warm.

Sixth week: Butcher meat birds.
Cull laying pullets.

Twelfth week: Move pullets to hen house.
Supply grit and oyster shell.

The Well Chicken

Chickens can suffer many different health problems from many different sources. With a little extra care from you, however, your chickens stand a good chance of staying healthy.

A Chicken's Lifetime

The average chicken lives between 10 and 15 years, although some have lived as long as 25 years. Few chickens live out their full, natural lives. Chickens raised for meat have a short life of only 8 to 12 weeks. Chickens raised for eggs or as breeders are usually kept for 2 or 3 years, until their production and fertility decline. Chickens raised for show are in their prime at 1 year of age. After that, prize winners may be shown for several more years or kept as breeders. Losers are usually culled.

The longer you keep a chicken, the more likely it is to get a disease. For that reason alone, many people won't keep chickens for more than a year or two. As soon as this year's chicks grow up, last year's flock is out the door.

Only chickens kept as pets have much chance of living a long life. But even a pet can be killed by dogs, get run over by a car, or come down with some disease.

A good fence prevents the first two disasters. Careful maintenance prevents disease.

Keeping Chickens Healthy

You already know many ways to keep your chickens healthy. One way is to start out with a healthy flock. Another is proper management of your flock.

- Ask your county Extension Service agent or state poultry specialist about necessary vaccinations in your area. If vaccinations are required, ask your Extension Service agent or a member of a local poultry club to show you how.

- At least once a day, give your chickens clean, fresh water. Scrub out their waterer once a week and rinse it in warm water with a squirt of chlorine bleach.

- Don't let old or moldy feed collect in troughs. If feed gets wet or moldy, or has manure in it, throw it away. Scrub the feeder and let it dry in the sun before refilling it. Throw away any wet or moldy feed still in the sack.

- Clean out your chicken coop at least once a year. Don't let piles of manure build up. Remove and replace wet litter. Keep old lumber and other junk away from your coop. Junk piles give disease-carrying mice and rats a place to hide.

- Provide enough space at feeders and waterers so the lowest chickens in peck order have room to eat and drink. Make sure your chickens have enough room to roam around.

- Spend at least 5 minutes each day talking to your chickens or singing to them. Showing your chickens you care helps keep them calm, free of stress, and less likely to catch a disease.

How Chicken Diseases Spread

Diseases can be carried through the air, soil, or water. They can be spread through contact with other chickens or other animals, especially rodents and wild birds. They may be carried on your clothing, especially your shoes.

Disease-causing organisms are always present in the environment. They may not cause problems unless a flock is stressed or its conditions are filthy.

Wild birds spread diseases by flying from one chicken flock to another, looking for spilled grain. If you live where there are lots of other chickens, place netting over your chicken yard to keep out freeloading birds.

Visiting other chicken yards is a good way to bring home diseases in manure clinging to your shoes. After a visit, clean your shoes thoroughly before tending your own flock.

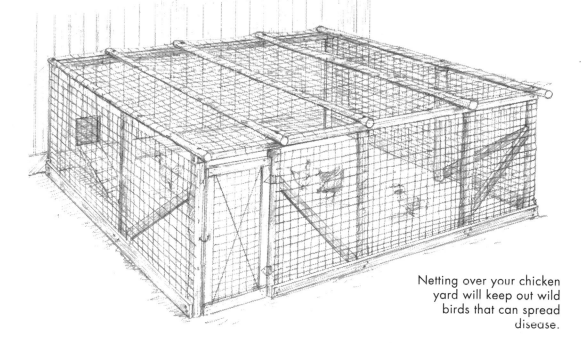

Netting over your chicken yard will keep out wild birds that can spread disease.

One chicken can get a disease from another, even if both birds appear to be healthy. After your flock is established, don't keep introducing new birds. Every time you bring home a new chicken, you run the risk of bringing home a disease. If you do acquire a new chicken, or bring one of your chickens back from a show, house it apart from the rest of your flock for at least 2 weeks, until you are sure the bird is healthy.

Precautions When Showing

Some shows require you to get a health certificate from a veterinarian before entering your chickens. Unfortunately, even health certificates do not mean all the birds in a show are healthy. A chicken that is capable of spreading a disease may not look sick to the vet who signed the certificate.

Your chickens are less likely to catch a disease if you take precautions to reduce stress. Start by training your chickens before the show. A chicken that is handled often and is used to being around people is under less stress than a chicken thrown into a carrier and hauled to a show without preparation.

Make sure your chickens don't run out of feed and water at the show or while traveling to and from the show.

If a chicken does catch something at a show, it will probably be a cold. Colds are spread among chickens the same way they are spread among people — by coughing and sneezing.

When you arrive at a show, look at the entries nearest yours. If a chicken doesn't look healthy or if one coughs and sneezes, bring it to the attention of the show's officials.

Happily, your chickens are not likely to catch a disease at a show. Most people who show take pride in keeping their chickens healthy.

Do not put medications in drinking water during a show. Medications cause a bird to drink less than usual, increasing stress. Besides, you can't tell what, if any, disease your chickens might be exposed to, so how do you know which medication to use?

How to Tell a Chicken Is Sick

The best way to tell a chicken is sick is to know how a healthy chicken looks and acts.

Sound. When your flock is healthy, you will hear your chickens "sing." A sick chicken may sneeze, gulp, or make whistling or rattling sounds when it breathes, especially at night.

Smell. Notice how your chicken house usually smells. Any change in odor may be a sign of disease.

Signs of Healthy and Diseased Chickens

Healthy	*Diseased*
Bright, full comb and wattles	Shrunken or off-color comb and wattles
Bright, alert eyes	Dull, closed, swollen, or watery eyes
Clean nostrils	Caked, crusted, or leaking nostrils
No breathing sounds, breathes through nostrils (except in hot weather)	Sneezes, gurgles, wheezes, rattles, or makes other odd sounds
Erect posture, head and tail held high	Head tucked in, neck twisted back, droopy tail or wings
Full breast	Shrunken breast
Firm abdomen	Hard or "squishy" abdomen
Smooth, clean feathers	Ruffled or stained feathers
Frequent eating and drinking	Eats less than usual, drinks more than usual
Grayish droppings with white caps	White, green, yellow, or runny droppings

Chicken Mortality

Dead chickens are, of course, one sign of disease. But don't jump to hasty conclusions if one of your chickens dies. The normal death rate, called the *mortality rate,* for chickens is 5 percent per year. If you have twenty chickens in your flock and one dies for unexplained reasons, there is your 5 percent. Naturally, you will be upset, but don't worry unless more chickens die or your flock has additional signs of disease.

Appearance. A healthy chicken looks perky and alert. It has a bright, full, waxy comb, shiny feathers, and bright shiny eyes. A sick chicken's feathers may look dull and its comb may shrink or change color. Its eyes may get dull and sunken, or may swell shut.

A sick chicken's droppings may be loose or bloody. Sticky tears may ooze from the corners of its eyes. Its nostrils may drip or get caked up.

Behavior. A sick chicken hangs its head or hunches down, sometimes ruffling its feathers to get warm. It may eat or drink less than usual, or lay fewer eggs. It may lose weight if it is a mature bird, or stop gaining weight if it is young.

By knowing how your chickens normally act, you can easily detect changes. Each time you enter the coop, stand quietly for a few moments. When your chickens get used to your being there, they will go on about their business. Then you can tell whether or not they look and act normally.

When to Get Help

The problem with chicken diseases is that many of them look alike, making it hard to tell which disease a sick chicken may have. You cannot properly treat a chicken unless you know what made it sick.

If you do not know exactly what disease your chicken has, you may give it the wrong medication and make things worse instead of better. Therefore, seek help from an experienced poultry person in your area, from your county Extension Service agent, or from your state poultry specialist.

Definitely call your county agent or state specialist if several chickens suddenly get sick or die at the same time. Your flock may have a contagious disease, meaning the disease could easily spread to nearby flocks.

You might think of calling a veterinarian, but most vets know very little about chickens. Even when they do, the cost of diagnosing and treating a chicken is usually more than the price of a healthy new bird.

One kind of veterinarian who knows all about chicken diseases is a *poultry pathologist.* If several of your chickens get sick, you might take the worst ones to a pathologist. You won't get the chickens back, but you will find out how to treat the rest of your flock. Your county Extension Service agent can help you find the nearest poultry pathologist.

Treating a Sick Chicken

If one of your chickens gets sick, or if you even suspect it might be sick, isolate the bird immediately. House it in a separate area, away from the rest of your flock. When you tend your chickens, care for the healthy ones first so you won't carry the disease from the sick bird to the healthy ones.

As cruel as it may sound, the best way to treat many diseases is to do away with the sick chicken. Kill the chicken, or have someone kill it for you, and burn the body or bury it so deep no dog can dig it up. By getting rid of a sick chicken, you make sure the disease doesn't spread. Besides, by the time you notice a chicken has a disease, it is usually too sick to be cured.

Even if you succeed in curing a bird, it may still be a carrier, meaning it can infect other chickens with the disease it once had. In addition, a cured bird is rarely as good as it once was at laying eggs, breeding, or winning show prizes.

Eliminate diseased birds from your breeding flock to make future generations more resistant to disease. By working hard to prevent diseases, you won't have to worry about treating sick chickens.

Poisons and Other Harmful Things

Chickens learn about new things by pecking at them. In doing so, they may eat something harmful.

Pieces of glass and other sharp objects, especially shiny ones, attract a chicken's attention. Eating something small and sharp can cause internal injuries. Protect your chickens by picking up bits of broken glass, tacks, staples, or other small, sharp objects you find lying around.

Another thing chickens find interesting is the string from a feed sack. Maybe it looks like a long, thin worm. If a chicken partially swallows the string, it can't pull the string back out. Whenever you tear off the string to open a sack of feed, put it in your pocket. Dispose of it properly, or wrap it around an empty toilet paper tube and save it. You never know when a piece of recycled string might come in handy.

Another harmful thing chickens swallow is pesticide. If someone sprays weeds or grass near your coop, make sure the pesticide is not harmful to chickens or else keep your chickens away.

Avoid spraying against cockroaches, wasps, or other insects near your chickens. Chickens eat dead bugs, and a poisoned bug can poison your chickens.

Parasites

Parasites are a common problem among chickens, but a problem you can easily treat. A parasite is an organism that derives benefit from another living thing without giving back anything in return.

Parasites can be brought to your flock by wild birds, rodents, and new chickens. They can also be carried by used feeders, waterers, nests, and other equipment. If you recycle used equipment, scrub and disinfect it before putting it in your own coop.

No-No's in the Chicken Yard

- No sharp objects
- No string
- No insect poisons

Two kinds of parasites affect chickens — internal and external. Internal parasites live inside a chicken, usually in its digestive tract. External parasites live on the outside of a chicken, either on its skin or on its feathers. Both cause stress and lower a chicken's resistance to disease.

Internal Parasites

Different kinds of internal parasites occur in different areas of the country. Your county Extension Service agent or state poultry specialist can tell you which kind are most likely to affect your chickens.

Coccidia. Coccidia are one kind of parasite that live everywhere. Coccidia cause coccidiosis. Although many different animals are affected by coccidiosis, the coccidia that affect chickens do not affect other kinds of animals. The reverse is also true.

Coccidiosis usually affects chicks (see page 92), but adult birds can be affected, especially in hot, humid weather. The first sign of coccidiosis is loose droppings, sometimes tinged with blood.

Medication for treating coccidiosis is sold through feed stores and poultry supply catalogs. Treat the whole flock at once. A properly managed flock develops natural immunity to coccidiosis.

Worms. Other kinds of internal parasites are commonly called worms. If you have a dog or cat, the worms you treat your pet for are similar to the worms chickens get, but not necessarily the same.

A chicken gets round worms by eating worm eggs. It gets tapeworms by eating an intermediary host infected with tapeworm. The intermediary host might be an earthworm, a grasshopper, a housefly, an ant, a snail, a slug, or any number of other crawly things.

Since confined chickens peck within a limited area, they are more likely than free-ranging birds to be infested with roundworms. Since free-ranged chickens eat more bugs than confined chickens, they are more likely to be infested with tapeworms.

Chicken eats intermediary host, becomes worm-infested

Worm eggs deposited with droppings

Intermediary host eats worm eggs

How chickens get worms

Being Sure About Worms

To find out for sure if your chickens have worms, scoop a few fresh droppings into a plastic bag, seal it, and take it to your vet for analysis. A vet looks for evidence of parasites through a microscope. Ask the vet to let you take a look, too. If your chickens have worms, the vet will tell you what kind and will recommend the proper treatment.

Symptoms of worms are droopiness, decreased laying, and weight loss, or, in young birds, slow or no weight gain. Loose droppings or diarrhea are also possible. Sometimes you may actually see worms in the droppings.

To reduce the chance your chickens will get worms, prevent puddles from forming in their yard and keep the floor of their house covered with dry litter. Do everything possible to keep wild birds and rodents away. Don't let junk pile up around your coop. Worm any new bird you bring into your flock.

External Parasites

Chickens discourage external parasites by taking dust baths. They dig holes, lie in them, and kick dust all over themselves. Afterwards, they shake off the dust and straighten up their feathers, one at a time. All this activity discourages external parasites from hanging around.

The main external parasites are lice and mites. They bite or chew a chicken's skin and suck its blood. If enough of them gang up on one chicken, they will kill it. Ask your county Extension Service agent for information on external parasites commonly occurring in your area, and what to do about them.

Check your chickens for parasites at least twice a month. Examine them at night, using a powerful flashlight to spot lice or mites. You won't need to check more than a few chickens, since external parasites spread rapidly from one to the other. Pick up one bird at a time and examine the feathers around its head, under its wings, and around its vent. Look carefully at the scales along its shanks, too.

Lice. If you see strings of tiny, light-colored eggs or clumps that look like tiny grains of rice clinging to feathers, you have found louse eggs. It is unlikely you will see the lice themselves, since they move and hide quickly. But you may see the scabs they leave on a chicken's skin.

Body mites. If you see tiny red or light brown insects that look like spiders crawling on the chicken's skin, you have found body mites. During the day, mites live on perches and nests. Because a setting hen stays on her nest day and night, she is a favorite target for mites.

Treatment for lice and body mites. When lice or body mites get into your flock, dust your birds with an insecticidal powder. Clean out your coop to discourage the parasites from coming back, and sprinkle insecticidal powder into all the cracks and crevices. Use only a powder approved for chickens, available through feed stores and poultry suppliers.

Leg mites. If the scales on a chicken's shanks are raised instead of lying smoothly, you have found leg mites. Leg mites burrow under the scales, forcing the scales away from the shank. Leg mites can be quite

raised scales

Raised scales on a chicken's legs indicate leg mites.

painful, causing a chicken to walk stiff-legged. To control leg mites, once a month brush perches and the legs of all your chickens with vegetable oil.

Feather Loss

Lice and mites can make a chicken's skin itch, causing the chicken to pick its own feathers and pull them out. If your chickens lose feathers because they are infested with mites or lice, you are not doing a good job of managing your flock.

Sometimes a rooster pulls out a hen's feathers during mating. A hen with no feathers on her back has no protection from the cock's claws and may be wounded. If your hens have bare backs, keep your rooster in a separate place and put him with the hens only one or two days a week or for only a few hours each day. If a hen is wounded, spray the wound with a veterinary wound spray.

Once a year, chickens naturally drop all their feathers and grow new ones. This process, called molting, is not a sign of external parasites or other disease. (For more about molting, see page 58.)

Winter Care

In cold weather, make sure your flock's drinking water doesn't freeze. Either use a livestock water heater or carry warm water from the house at least twice a day.

A cock's comb may freeze during the night. Unlike a hen, a cock doesn't sleep with his head tucked under his wing. Insulating your coop helps keep your chickens warm, as does mounting a small electric heater above the perch. To save electricity and to make sure your chickens won't get too warm, plug the heater into a thermostatic control that kicks on when the temperature falls below 35°F.

During winter, both cocks and hens suffer if their house is drafty. A chicken ruffles up its feathers to trap warm air next to its body. Wind blowing through the feathers removes the warm air, giving the chicken a chill. Check your coop to make sure it isn't drafty. Cover cracks and crevices to keep the wind from blowing through.

Summer Care

If the temperature reaches 95°F or above, your chickens may suffer from the heat. Make sure they have plenty of cool, fresh water. Keep the waterer in the shade or carry cool water from the house. Since chickens drink less when their water isn't pure, avoid putting medications in drinking water during hot weather.

In warm weather, chickens breathe through their mouths instead of through their nostrils. A mature chicken starts breathing through its mouth when the temperature reaches 85°F. A chick breathes through its mouth when the temperature is 100°F or more.

Breathing through its mouth, breathing rapidly, and spreading its wings away from its body are a chicken's way of keeping cool. They are also signs of heat stress. Even when a chicken does all these things, if the temperature gets to 105°F or above, the chicken may die.

In very hot weather, make sure your chickens have shade where they can get out of the sun without crowding together. If your climate is dry, keep your chickens cool by spraying them lightly with a hose. As the water evaporates, your chickens will feel cooler. (If your climate is humid, spraying won't help. When the air is already full of moisture, evaporation cannot occur.)

In warm weather, hens lay smaller eggs with thinner shells. Don't worry. As soon as cool weather comes, their eggs will get back to normal.

Health Maintenance Chart

Date	Medication Used	Dosage	Remarks

Show and Tell

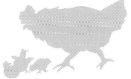

All kinds of contests and shows are held that relate to chickens. They are held for two reasons. One is to demonstrate your skills. The other is to have fun. Many contests and shows are also learning experiences. Sometimes they point the way toward careers in the egg or broiler industry. To find out about shows or contests in your area, contact the poultry superintendent for your county fair, your county Extension Service agent, or your state poultry specialist.

Look for programs sponsored by youth groups or that include a youth division. In a youth division, you compete against others with your same level of knowledge and skill. After you gain experience, challenge yourself by competing in an open division, where you can sharpen your skills by competing against experienced adults.

Contests

Flying Contests

Chicken-flying contests are a tradition among people who grow up in the country. One young fellow enjoyed this sport so much he started the International Chicken Flying Association (ICFA), which has many state, county, and local chapters. To find the chapter nearest you, contact the ICFA at the address listed in the back of this book.

In a chicken-flying contest, each chicken must fly down from a 10-foot post. On top of the post is a mailbox with both ends removed. A chicken placed in the mailbox can fly in only one direction — out the front. Its flying distance is measured from the base of the post to where the chicken first touches ground.

Crowing Contests

Rooster-crowing is another contest held by the ICFA. Each rooster has 1 minute in which to sound off. He is judged by the number of times he crows, the intensity of his crowing, his determination, his tone, and his crowd appeal. The cocks' owners compete, too, by trying to outcrow each other.

Egg-Laying Contests

Egg-laying contests test your skills at raising and selecting good laying hens. Hens are not judged on the actual number of eggs they lay, but on their probable ability to lay based on their appearance, as described in Chapter 5.

Hens will likely compete in "pens" of three or four. Enter hens that look as much alike as possible. This is not a beauty contest but a production contest. Your best layers will have worn, frayed feathers, not the smooth, sleek feathers of an exhibition hen.

Production Hen Judging Contests

Another contest related to laying hens tests *your* ability to judge. Learning to judge production hens teaches you to become a better manager of your own laying flock.

You might compete alone or you might join a team. After looking over a group of hens and deciding which is the best layer, which is the second best, and so on, you must explain why you placed each hen as you did.

Long-distance Flyers

Although it is claimed that a barnyard bantam named Sheena holds the world's record for flying 630 feet 2 inches, the official ICFA record is held by a barnyard bantam named Judy, who flew 542 feet 9 inches in 1989. The ICFA offers a big prize for any chicken that can beat Judy's record.

You are judged on your appearance and confidence, the accuracy and completeness of your descriptions, and your use of correct language. You must know the proper words for the parts of a hen's body and for production characteristics.

Show Egg Contests

Some contests determine who has the best show eggs. While there are many variations, the most common version requires you to enter 1 dozen eggs. The eggs should be as nearly perfect as possible and alike in shape, size, color, shell texture, and interior quality.

In preparing for this contest, gather eggs four or five times a day. Select only eggs that are spotlessly clean and free of shell defects. Select oval eggs with one end larger than the other, never round or oblong eggs.

Use a kitchen scale or postal scale to weigh each egg to find the closest match and to be sure your eggs are the right weight for the class you are entering. Candle the eggs and eliminate those with double yolks, blood or meat spots, or cracked shells. Cracks show up during candling as thin white lines.

Store your show eggs in the refrigerator to protect their inner qualities. At show time, wrap the eggs in a towel and carry them to the show in an insulated picnic carrier.

Egg-Grading Contests

In an egg-grading contest, you judge eggs based on their exterior and interior qualities. Exterior qualities are those pertaining to the shell: shape, texture, stains, dirt, and ridges or wrinkles (occurring when a shell cracks and seals back up before it is laid).

Interior qualities include air cell size, firmness of white, outline of yolk, and the presence of blood or meat spots. Entering egg-grading contests helps you learn more about egg quality.

The Boston Poultry Show

The first poultry show in the United States was held at the Public Gardens in Boston, Massachusetts, in 1849. Poultry shows soon became popular, encouraging people to perfect existing breeds and develop new ones. Many of the unique breeds we have today might not exist if it had not been for that first show.

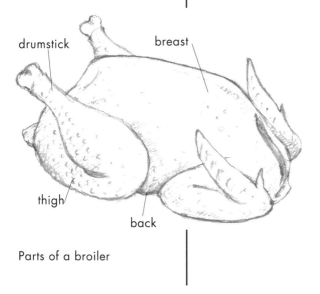

drumstick

breast

thigh

back

Parts of a broiler

Egg-Cooking Contests

Another contest pertaining to eggs is the national egg-cooking contest. You must win a state division to qualify for the national cook-off.

Recipes are judged on creativity, taste, appearance, ease of preparation, and nutritional value. This contest is sponsored by the American Egg Board. Its address is at the back of this book.

Meat-Production Contests

Meat birds, like laying hens, are usually shown in pens of two, three, or four. Select birds that look as much alike as possible and that are similar in weight. Meat chickens are judged on

- Fleshing (meatiness of the breast, thigh, and drumstick)
- Conformation or body structure (as close as possible to the ideal rectangular shape)
- Finish (creamy or yellow in color, not reddish or bluish)
- Feathering (mature feathers only, no pin feathers)
- Freedom from defects (no crooked breast bones, bruises, cuts, tears, callouses, or breast blisters)

When selecting chickens for a meat-production contest, try to picture what they would look like dressed. A good way to learn what dressed chickens should look like is to enter a meat-grading contest. You will judge dressed broilers or roasters based on their shape, condition, and degree of meatiness.

Chicken-Cooking Contests

Another contest related to meat birds is the chicken-cooking contest. In addition to entering local contests, try your luck at the competition sponsored by the National Broiler Council. (The address is at the back of this book.) Your recipe will be judged on taste, appearance, appeal, and simplicity.

Avian Bowl

The Avian Bowl tests your overall knowledge about chickens and other birds. It works like the TV quiz show, "Jeopardy." Local winners compete in a statewide contest. Each year's state winners compete in a national play-off. To find out if there is an Avian Bowl in your area (or to help start one), contact your county Extension Service agent. An Avian Bowl study packet is available from the address listed at the back of this book.

Poultry Shows

Exhibition chickens are the kind you are most likely to see if you visit the poultry show at your county fair. Poultry shows are sponsored by fair boards, the APA, the ABA, and various local clubs.

Poultry is a word that describes chickens grouped together with other domesticated food-producing birds such as turkeys, ducks, and geese. Another word for poultry is *fowl.* Ironically, the fowl you find at most poultry shows are bred for beauty rather than food-producing abilities. Since they are bred specifically for exhibition, they are sometimes called *exhibition fowl.* Since they are bred according to the *American Standard of Perfection* or the *Bantam Standard,* they are also called *standard-bred fowl.*

A show may be *sanctioned* by either the APA or the ABA. The sanctioning organization allows its name to

How chicken meat is graded

Grade A:
Sides of breast curve out

Breast bone is well fleshed.

Grade B:
Sides of breast are flat

Breast bone is fairly well fleshed.

Grade C:
Sides of breast curve in

Prominent breast bone, poorly fleshed.

be used in connection with the show and offers prizes. When selecting birds for a sanctioned show, use the standard published by the sanctioning organization. For an APA-sanctioned show, use the *American Standard of Perfection*. For an ABA-sanctioned show, use the *Bantam Standard*.

Prepare yourself to show by visiting poultry shows in your area. Notice the quality of the birds on display and the condition they are in. Ask the show secretary or superintendent for a copy of the show's *premium list*.

The Premium List

The premium list tells you what organization, if any, has sanctioned the show and what prizes are offered. Prizes are also called "premiums," which is how the premium list got its name. Typical premiums include ribbons, trophies, and small amounts of cash.

The premium list tells you the deadline for sending in your entries, the cost for entering each bird, and when and where the show will be held. It gives the rules for entering and the requirements for each *class*. A class is a group of chickens judged against each other, organized according to *Standard* classifications.

The chickens you enter into a show must fit one of the classes named in the premium list. Not only is it embarrassing to enter a chicken in the wrong class, but your entry may be disqualified from the show.

Each bird must also be correctly entered according to its breed and variety, and whether it is a cock, hen, cockerel, or pullet. A bird may also be entered as a part of a *trio* (one male and two females of the same age, breed, and variety) or a *pen* (three or four birds of the same age, sex, breed, and variety).

Preparing for a Show

The premium list tells you whether or not any vaccinations, blood tests, or health certificates are required. Show only healthy birds. To do otherwise is unfair to the other exhibitors and to your chickens. Make sure your birds are free of lice and mites. Some judges refuse to handle infested birds.

Well before the show, separate your males from your females and the males from each other. Males may fight and injure each other. Allow enough time for injuries that have already occurred to heal. Males may break a female's feathers during breeding. Pull out broken feathers in plenty of time for them to grow back.

Examine each bird to make sure it won't be disqualified. The *Standard* lists all the possible reasons for disqualification. Some apply to all chickens, some to specific breeds.

General disqualifications include a crooked back, a wry or crooked tail, and *stubs* (downy feathers on the shanks or toes of a clean-legged breed). Such defects

Classes

American	Buckeye, Chantecler, Delaware, Dominique, Holland, Java, Jersey Giant, Lamona, New Hampshire, Plymouth Rock, Rhode Island Red, Rhode Island White, Wyandotte
Asiatic	Brahma, Cochin, Langshan
English	Australorp, Cornish, Dorking, Orpington, Red Cap, Sussex
Continental	Campine, Crevecoeur, Faverolle, Hamburg, Houdan, La Fleche, Lakenvelder, Polish
Mediterranean	Andalusian, Ancona, Catalana, Leghorn, Minorca, Sicilian Buttercup, Spanish
Other	Ameraucana, Araucana, Aseel, Cubalaya, Frizzle, Malay, Modern Game, Naked Neck, Old English Game, Phoenix, Shamo, Sultan, Sumatra, Yokohama

Bantam Classes

Single comb, clean-legged	Any standard-bred bantam with a single comb and no feathers on its shanks
Rose comb, clean-legged	Any standard-bred bantam with a rose comb and no feathers on it shanks
Any other comb, clean-legged	Any standard-bred bantam with a comb other than a single or rose comb and no feathers on its shanks
Feather-legged	A standard-bred bantam with any kind of comb and with feathers on its shanks
Game	Any standard-bred Old English or Modern Game bantam

are important because they are almost always passed on to a chicken's offspring.

If you plan to show Modern or Old English Games, your cocks must have their combs and wattles trimmed, or dubbed, in order to qualify. Exhibitors who don't like to *dub* or who live in states where dubbing is illegal show only cockerels, hens, and pullets. To find out more about dubbing, contact a Game breeder in your area.

After deciding which of your birds to show, spend time training and conditioning them. Training involves teaching a bird how to act while on display. Conditioning involves making the bird look its best.

Training

Training your chickens before each show prepares them for what they will experience during judging at the show. A frightened bird may crouch in its cage. It may fly up when people come near. It may struggle when the judge handles it. It's hard for a judge to tell what a frightened bird really looks like, so a calm bird will usually place higher than a frightened one.

At least 3 weeks before a show, isolate each bird in a separate cage, similar to those used at shows. A show cage is called a *coop*. When you are told what time to "coop in" or "coop out," you know what time you must place your birds in, or remove them from, their assigned coops.

While your potential prize-winners are in their training coops, handle each bird two or three times a day. Begin by approaching the coop slowly and opening the door. If the bird seems frightened, pause until it calms down.

Reach across the bird's back and place one hand over its far wing, at the shoulder. Place your other hand under the bird's breast, with one leg between your thumb and index finger and the other between your second and third finger. Your index finger and second finger should be between the bird's legs. The bird's breastbone, or keel, will rest against your palm.

Gently lift the bird out of its coop, head first. Hold the bird quietly for a moment, then remove your hand from its wing. Let the bird sit in your hand another moment. Turn it to examine its comb and wattles. Open each wing. Always keep your bird facing toward you. If you face it away from you, it may try to escape.

Gently return the bird to its coop, head first. Place it on its feet, let go, and slowly close the door.

Sometimes, no matter how gentle and patient you are, a bird doesn't want to be handled. If, after a week or two, a bird still acts frightened, do not enter it this time. If you wish, try to train it for the next show.

Washing Your Chickens

Just as you take a bath and spruce up before going somewhere special, your chickens need to be washed and spruced up before a show. Until you get the hang

Safe Handling

Always remove a bird from its coop, and replace it, head first so its wing feathers won't get caught in the doorway and become damaged.

Drying Time

Wash your chickens 48 hours before a show. It takes only 20 minutes to wash a chicken, but it takes 18 hours for a bird to dry and 48 hours for its feathers to get back their natural oil.

Do not substitute bleach for bluing! It can be harmful to your chickens.

of washing chickens, practice on a few birds you don't plan to show.

Start by filling a washtub or basin with warm water. The water is the right temperature if you can comfortably hold your elbow in it for one minute. Add enough liquid soap to make suds. Do not use a harsh detergent, which makes feathers brittle.

Bathe one chicken at a time. Place one hand against each wing, so the bird can't flap, and immerse it to its neck. If the bird struggles, dip its head under water for an instant and it will usually calm down.

Use a sponge to soak the feathers through to the skin. To avoid breaking feathers, rub only in the direction they grow. Use an old toothbrush to scrub dirt from the bird's shanks and toes. If the chicken won't be still for this, wrap it in a towel so that only its legs and feet stick out. This keeps it from flapping its wings while you clean its shanks and toes.

When the chicken is clean, soak it for 2 minutes in fresh warm water, slightly cooler than the wash water. Work out the soap by moving the bird back and forth in the water.

Rinse the bird a second time. If it has white feathers, add two drops of liquid laundry bluing — if you can find it in your area — to make them whiter. Take care not to add too much or your chicken may turn blue!

Mix a little rubbing alcohol with an equal amount of water. Dip a soft cloth in the mixture and use it to clean the bird's comb and wattles, *taking care not to get any in the bird's eyes*. Rub the shanks and comb with Vaseline or baby oil, taking care not to get oil on the feathers.

Clean beneath the bird's toenails with a toothpick. Trim long nails with nail scissors or clippers. Trim a little at a time, being careful not to cut too much and cause bleeding.

If the temperature outside is at least 70°F and it isn't windy, dry the bird outdoors. To speed things up, some people use a hair dryer, but your chicken will look much better if you let it dry naturally.

Dirt that touches wet feathers may stick, so put the bird in a coop or a pet carrier lined with fresh litter until it dries. Place one bird in each cage. Crowded birds may injure each other or soil each other's feathers.

At the show, just before the judging, give your bird a final grooming. Wipe off any dirt sticking to its feathers. Reclean its shanks, feet, face, and comb. Rub a tiny amount of baby oil or Vaseline on its shanks, feet, and comb with a soft cloth. (If you use too much Vaseline or oil, dust will stick to your bird's comb and feet.) Shine up its feathers with your hands or with a piece of silk or wool.

How the Judging Works

Each show entry is judged in these three ways:

- Is the bird appropriate for its class?
- Is it in good condition?
- How does it compare with the others in its class?

A judge licensed by the American Poultry Association is supposed to judge birds according to the scale of points in the *American Standard of Perfection*. A certain number of points is given for each trait, such as comb, tail, back, and so forth. Since no bird can be perfect, no bird ever gets a perfect score of 100 points.

Before the judging starts, compare all the entries in your class and decide how you would place them if you were the judge. When the judging is over, compare your placings with the judge's. If your bird did not place first, try to find out what makes the other birds better than yours.

For some youth shows, you may be asked to bring your bird to the judge's table. Not only is the bird judged for its appearance and condition, but you are judged for your appearance, conduct, knowledge of poultry, and ability to show and handle your bird.

At most shows, however, the judge looks at birds individually in their coops. Some judges explain their reasons for placing each bird. Others do not. Some judges fill out evaluation cards. Others do not.

After the judging, ask the judge to talk to you about your bird's good and bad qualities. Never ask a judge to discuss your bird until after its class has been judged. If

Show Record

Date	Name of Show	Bird Shown	Number in Class	Placing	Premium	Entry Fee	Judge	Remarks

the judge doesn't have time to talk to you, ask one of the show officials to point out an experienced poultry person willing to answer your questions.

In discussing your bird with the judge or with other exhibitors, don't take criticism personally. Use it to improve your next entry. Even if your bird wins a blue ribbon, gather ideas for future improvement. You never know when you might run up against tougher competition at the next show.

Whether or not you bring home a prize, remember that what you learn at a show is more important than what you win.

Backyard Contests

Here are two contests to organize among your friends.

Egg Toss

This contest is designed to see how far teams of two can toss and catch a raw egg without breaking it. Hold the contest in an open field where broken eggs won't create a problem.

Team members face each other in two rows. The members in one row toss their eggs to their teammates in the other row. Any team that breaks its egg is out. Those who catch their eggs take a step backwards and toss again. The last team with an unbroken egg wins.

Egg Tapping

This contest is designed to see whose eggs have the hardest shells. Each contestant starts with one dozen hard-boiled eggs. Everyone sits in a circle with an egg in the palm of each hand, small end up.

Tap the egg in your left hand against the egg held by the person on your left, while the person on your right taps your other egg. When the small end of an egg cracks, turn the big end up. When both ends crack, take a fresh egg from your original dozen.

Contestants who crack all their eggs are out. The last person with an uncracked egg wins. (Here is the trick: Protect all but the tip of your egg with your fingers.)

Eggonomics

Chickens offer so many different ways to make money, you may soon find yourself running a business whether you intended to or not. Even if you don't wish to sell chickens or their products, you have to budget your money so you always have enough to buy feed.

Start-Up Costs

Now that you know all the things you need to get started raising chickens, you are ready to prepare a start-up cost analysis. Begin by listing all the items you need. As a guideline, the accompanying chart lists most items and their average prices at the time this book was written.

Fill in the "estimate" column based on information from poultry supply catalogs and from your local feed store. If you already have one of the items listed, or you can get it for free, place a "0" in that column.

Include a little extra under "miscellaneous" for things you will need that you might not think of in advance. For example, a waterer might break and need to be replaced.

After you fill in all the blanks, add up the "estimate" column to get your total start-up cost. If it seems too high, look for ways to reduce costs, starting with the most expensive items. For example, instead of con-structing a new hen house, do you have a building you can fix up?

Start-Up Cost Analysis

	Average	Estimate	Actual
1. Housing (8' x 12')	340.00		
Nests (6)	16.00		
Roosts	1.75		
2. Fence	.60/ft		
Gate	10.00		
3. Lighting			
Timer	17.60 ea		
4. Waterer	3.80 ea		
Chick waterer	1.46 ea		
5. Feeder	12.40 ea		
Chain	1.00/4 ft		
6. Feed, starter	7.50/50#		
7. Feed storage bin	8.50 ea		
8. Bedding	1.25		
9. Isolation cage	10.00		
10. Medications			
11. Numbered leg bands	3.80/24		
12. Chicks	30.00/25		
13. Miscellaneous			
14. **Total**			

As you acquire each item, list the price you actually pay in the column marked "actual." When you add up the "actual" column, compare the total with your initial estimate. How did you do?

Financing Start-Up

Getting money to finance start-up is a problem every business faces. The most common reason adults fail in business is that they do not have enough money to get a proper start. They underestimate their costs and overestimate their income potential.

If you run your flock like a business, you will learn a valuable, lifelong lesson. By working out your finances ahead of time, you won't have to keep worrying about getting money to support your chickens before they are ready to support themselves.

If you raise broilers, set aside enough to pay for feed to the end of the project. If you raise laying hens, set aside enough for feed until they start laying full-size eggs.

If you raise exhibition fowl, income will be farther down the road than with broilers or layers. You may earn a little in prize money, but most of your income will come from the sale of chicks or show birds you have raised. Not all of your birds will be good enough to show or breed. The less desirable ones can be used for food.

You might finance start-up costs from your savings, from money you earn doing odd jobs such as mowing lawns or washing cars, or by getting a loan from a relative. If you arrange a loan, be ready to show how you can pay it back through the sale of products from your flock. Perhaps you can repay part of the loan in eggs or dressed broilers, saving your cash income to pay maintenance costs. Each time you make a delivery, deduct the value of the eggs or broilers from the loan balance.

Maintenance Costs

The biggest maintenance expense for your chickens is the cost of feed. Other maintenance costs include bedding and medications. You might also have to pay for electricity (if you light your hen house) and rent (if you keep your chickens on someone else's land).

If you show your chickens, you must pay entry fees. To qualify for certain premiums, you may wish to join the APA, ABA, or some other organization, for which you must pay annual dues.

Estimating maintenance costs in advance helps you budget your money so you never run short. Keeping track of your costs lets you know whether you are running your business at a profit or at a loss.

Profits and Losses

Keep track of your income and expenses in a bound ledger book. At the left of each entry, write the date. Then, explain the source of the income or expense. On the right side of the page, make one column for income and one for expenses. List each amount in its appropriate column.

Make your entries neatly. Rather than writing in your ledger as income and expenses occur, you may prefer to use a sheet of notepaper. Transfer your notes to your ledger at a quiet time when you can concentrate on neatness.

Instead of keeping a handwritten ledger, you might keep track of your income and expenses with a personal computer and accounting or spreadsheet software. Accounting software is inexpensive and simple to use, and lets you create charts and graphs to show how your business is doing.

Running a Business

Every month, add up the income column and the expense column of your ledger. Some months the income column will be greater, showing that you earned a profit. Other months the expense column will be greater, showing that you took a loss.

If you consistently take a loss, you do not have a business but a hobby. There is nothing wrong with that. Some people spend their money on model airplanes or on video games. Others have more fun raising

Income and Expenses Ledger

Date	Entry	Income	Expenses

chickens. With *your* hobby, you learn how to handle animals. You also have the satisfaction of producing wholesome eggs and meat for yourself and your family. That is something you cannot put a price tag on.

If you wish to run a business, you must try to earn a profit each month. Not every business succeeds in earning a profit every month, especially at the beginning. When your business starts earning profit, don't be too quick to spend it at the mall. Set some money aside for maintenance costs or to improve or expand your business.

If you borrowed money to finance your business, use part of your profit to pay back the loan. If you repay your loan in 3 years or less, you are doing better than most adults who start a business.

Sources of Income

Selling Eggs

The most obvious way to earn money with your chickens is by selling eggs. Nearly everyone who raises chickens has extra eggs at one time or another.

Prices for commercially produced eggs fluctuate throughout the year. The price goes up when feed costs and consumer demand go up. The price goes down when feed costs and consumer demand drop. Your egg customers will expect the price of your eggs to fluctuate with the market.

Decide how much to ask for your eggs by visiting three grocery stores at least once a month. Note the price of eggs at each store. Add up the three prices and divide by three to get an average price. This average price is the least you should ask for your eggs. Since *your* eggs are fresher and tastier than store-bought eggs, sell your eggs for a few cents more.

Eggs sold at the market are sorted according to size. You likely won't have enough eggs to sort strictly by

Egg Marketing

Price /Doz.	Jan	Feb	Mar	Apr	May	Jun	Jul	Aug	Sep	Oct	Nov	Dec	Doz. Laid/M
1.20													70
1.15													65
1.10													60
1.05													55
1.00													50
.95													45
.90													40
.85													35
.80													30
.75													25
.70													20
.65													15
.60													10
.55													5
.50													0

Each month, mark an "x" to represent the average market price for a dozen eggs. Mark an "o" to represent the average number of eggs, in dozens, your hens laid that month. (Divide total dozens for the month by the number of hens in your flock.) At the end of the year, draw one line to connect all the x's, and another line to connect all the o's. Can you see a pattern?

size, so chances are you will sell mixed sizes. Price them according to the smallest eggs in the carton. For example, if you package large and medium eggs together, sell them for the price of medium eggs. Set aside especially small or especially large eggs for your own use.

Getting Egg Cartons

Getting enough egg cartons is no problem. Simply set up your own recycling program. Ask your family and friends to save empty egg cartons for you. Egg cartons are easy to store if you open them out and stack one on top of the other.

Sort through your recycled cartons and eliminate torn or dirty ones. If you have more than one kind of carton, place a few eggs in each and see which carton makes your eggs look best. Some styles make your eggs look small, others make them look large.

Covers sometimes have openings in the top so customers can see the eggs without opening the carton. Since holes hasten evaporation, cartons without holes keep eggs fresher longer.

Selling Laying Hens

Perhaps you have more laying hens than you need and you wish to sell a few. If there is a demand in your area, you might raise extra hens on purpose to sell. Your market will be customers who want a few layers, but don't want the bother of raising them from chicks.

You will get the best price for pullets just starting to lay. Any customer, no matter how inexperienced, can easily tell the birds are young and still have a long life ahead of them.

Year-old hens sell for somewhat less than pullets. If you have a thriving egg business, you might wish to raise a new flock each year to keep production high. Sell the old hens as soon as the pullets start to lay. The older hens will still lay well enough to bring a reasonable price.

How much to charge for hens or pullets depends on two things: how much hens in your area sell for and how much the birds cost you to raise. To get an idea of local prices, look for classified ads in the newspaper, ask at the feed store, and call your county Extension Service agent.

You will know how much each bird cost you to raise if you keep careful records. When setting a price on older hens, remember they have already partially paid their way by giving you eggs to sell.

You may earn a dandy profit selling hens or pullets. On the other hand, chickens may be so cheap in your area that it doesn't pay to raise them for sale. Study the market before you go into the hen or pullet business.

Selling Broilers

Selling broilers can be tricky business. In some states you need a license to sell dressed broilers. Find out about your state's laws by contacting your state poultry specialist.

Even if you can legally sell dressed broilers in your state, you may have a hard time competing against low grocery store prices unless your customers are willing to pay extra for homegrown chicken. Customers expect perfection, though, so take great care in dressing and packaging your broilers.

If you live in a rural community, you might avoid dressing your broilers by selling live birds to customers willing to dress their own. To make sure you have a customer, arrange the sale in advance, preferably before you buy or hatch the chicks.

When setting a price on your broilers, keep careful records of your costs and add a little for profit. Remember, each broiler requires at least 7 pounds of starter to reach 3½ pounds. To estimate your cost in advance, multiply the number of broilers you wish to raise times 7 pounds of feed times the price of feed per pound. Add 10 percent in case the cost of feed goes up or your broilers eat more than you expected.

In marketing broilers, the idea is to grow your birds to eating size in the shortest possible time at the least possible cost. To turn a profit, your management must be nearly perfect.

Selling Chicks

Because you don't have to put a lot of money into feed, your cost is less if you sell newly hatched chicks. On the other hand, chicks sell for considerably less than grown birds. In addition, many states require breeder flocks to be tested for a disease called *pullorum* before their chicks can be sold.

Chicks sell best in spring, around Easter time. But do not advertise "Easter chicks" for sale. In some areas, doing so is illegal. Too many people buy Easter chicks because they are "so cute." When the novelty wears off, they leave the chicks to suffer and die.

Sell chicks only to people like you, willing to take time to learn how to do it right. Whether or not you make money depends on the demand for chicks in your area. Are customers interested in buying chicks at all? If so, are they interested in buying the kind you have?

If you are very sure of the market, hatch the chicks, then run an ad in your local newspaper or post notices on community bulletin boards. If you misjudge the market, however, you may get stuck with chicks no one wants.

Instead, you might arrange in advance to sell your chicks to a local feed store. You must give the store a discount, because the feed store has to sell the chicks at a higher price in order to make a profit. But even though your selling price is lower, you will have a sure sale.

Agree to an advance sale only after you have practice operating your incubator and you are sure you'll be able to deliver. Otherwise, like the boy who cried wolf, if you've failed to deliver once, the next time you try to arrange an advance sale your potential customers may not believe you can deliver.

Selling Hatching Eggs

The price you can get for hatching eggs is higher than for eating eggs, but lower than for newly hatched chicks. In addition, in most parts of the country demand is quite small. You may, however, find a customer with a setting hen or a new incubator who is looking for fertile eggs to hatch.

Selling hatching eggs saves you the trouble of incubating the eggs yourself. It isn't fair, however, to sell hatching eggs unless you have hatched a few and know from experience that your eggs are fertile and hatch well. In many states, you must also have your breeder flock tested for the disease pullorum before you can sell eggs for hatching.

When selling hatching eggs, the only guarantees you can offer are that the eggs come from a certain breed and that they have been properly stored. You cannot guarantee the eggs will hatch because you have no control over how they are incubated. Customers may blame you for selling eggs that don't hatch, even though it isn't your fault.

Selling Show Chickens

How much you get for show birds varies widely with locale, breed, and quality. The only way to set a price on a particular bird is to find out how much comparable birds sell for.

Whether or not you find a market for show breeds depends on how well you do at shows. If your birds consistently lose, you won't be able to give them away. If your birds consistently win, customers will beat a path to your door.

Naturally, you wouldn't sell your best chickens. You need them to keep your business going. But don't sell culls, either, or they will ruin your reputation. The birds you sell are your best advertising. If you sell culls and your customers enter these poor-quality birds in shows, everyone will know where the birds came from. If your customers show winners, everyone will know that, too.

When a bird you sold beats your own entry at a show, be proud. It is the best advertising you could ask for.

Selling Manure

Every year, each chicken produces approximately 25 pounds of manure. Part of it gets trampled into the ground as your chickens wander around during the day, but a large portion is deposited beneath the roosts at night.

Chicken manure is prized by organic gardeners as fertilizer for vegetables and flowers. But, due to its high nitrogen content, fresh chicken manure can "burn" plants. Gardeners therefore prefer to get fresh manure in the fall so it has all winter to decompose before spring planting.

Instead of selling fresh manure, you might compost it. Gardeners pay more for composted manure because it has no odor and can be used any time. Composting is not difficult. Directions can be found in any good book on organic gardening.

How much you get for your chicken manure depends on how much customers are willing to pay and on how much work you have to do. Charge more for composted manure than for fresh manure. Charge more if you deliver the manure than if your customer picks it up. If your customer helps you clean out your coop, you might trade free manure for the labor.

Selling Feathers

Chicken feathers are used to tie flies for fishing lures, to make jewelry and home decorations, and to trim hats and other clothing. Different crafts require different kinds of feathers, so research the market carefully before going into the feather business.

Not every feather of every chicken is valuable. For fly-tying, only a cockerel's hackle and saddle feathers are used. Other crafts use wing and tail feathers.

Some breeds produce more valuable feathers than others. The best feathers for fly-tying, for example, come from fast-growing, hard-feathered breeds in colorful varieties. Examples are Barred Plymouth Rock, Blue Andalusian, Buff Minorca, and Silver Penciled Wyandotte. Many markets prefer the smaller feathers of bantams. If you are interested in feathers for fly-tying, be sure to start with a strain developed for that purpose.

A chicken must be killed before its feathers can be harvested. Hackle feathers are sold in *capes,* with the dried skin attached. Saddle feathers are sometimes sold individually, but are more valuable as a saddle patch, with the dried skin attached. Wing and tail feathers are sold individually, one dozen or more to the package.

You can make quite a bit of money selling feathers, especially for fly-tying. But you must have the right feathers to suit the buyers' needs. Furthermore, your feathers must be properly harvested, processed, and packaged.

To learn more about selling feathers, visit craft stores and fishing tackle shops, study fishing supply catalogs, and look up fly-tying and hobby books in the library. Get involved in fly-tying or other feather crafts to learn, firsthand, what types of feathers are needed, how much they are worth, and how to market them.

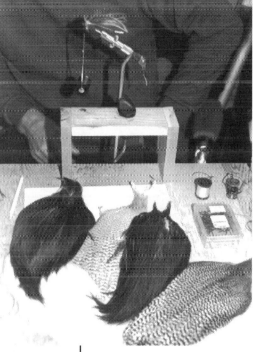

Hackle feathers for fly tying are sold in capes.

Dealing with Customers

Always be courteous when dealing with customers. Be prepared to answer questions. Customers who buy eggs or broilers may ask how they were grown. Customers who buy chicks or grown birds may ask advice on feeding or housing.

Be prepared to make change when necessary. Keep an assortment of coins and 1- and 5-dollar bills handy so your customers won't have to wait while you look for change.

If customers come to your house, have everything clean and ready. Don't expect a customer to wait, for example, while you look for an egg carton and run out to the coop to collect eggs. Imagine how you would feel if the clerk at a store rummaged around every time you went to purchase something.

Remember, above all, you are running a business. You have full control over the quality of the products you sell. Whether those products are inferior or superior, word will get around. Sell only your best, and you will soon have a reputation for being a wonderful person to do business with.

Flock Maintenance Summary

Daily

Feed your flock twice a day.
Provide clean water twice a day.
Replace wet litter.
Collect eggs and note the number on your record sheet.
Check the health of your flock.
Check your fence for holes or other problems.
Record income and expenses.

Weekly

Scrub all feeders and waterers.
Add litter to nests, if needed.
Check for mites and lice every 2 weeks.

Monthly

Add litter to coop floor.
Tally monthly income, expenses, and egg records.
Brush perches and legs with vegetable oil.

Winter

Collect eggs often so they won't freeze.
Give your flock warm water twice daily.

Provide layers with 14 hours of light.
Check coop for drafts.
Watch for frozen combs in cold weather.

Spring

Hatch chicks.
Clean and disinfect coop.
Eliminate puddles where spring rain collects.
Set mouse traps near feed storage bins.
Keep your flock away from sprayed lawns or gardens.

Summer

Collect eggs often so they won't spoil.
Make sure your flock has plenty of shade.
Provide cool water; keep it in the shade.
Cull undesirable chicks.
Keep hens and broilers away from onions and garlic.

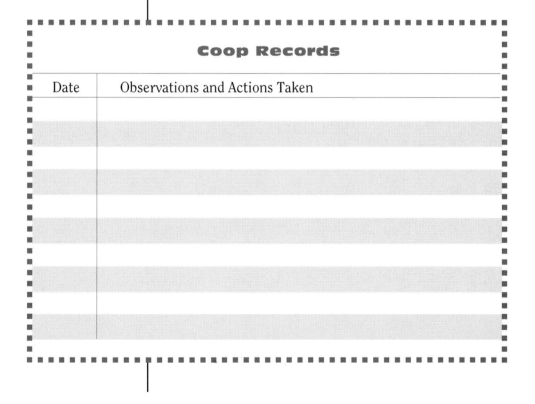

Coop Records

Date	Observations and Actions Taken

Helpful Sources

State Poultry Specialists

The following offices are listed in alphabetical order by state. To contact your state's poultry specialist, address your letter to: Extension Poultry Specialist, Cooperative Extension Service.

Example:
Extension Poultry Specialist
Cooperative Extension
  Service
University of Tennessee
Knoxville, Tennessee 37901

Alabama
Auburn University
Auburn, AL 36849

Alaska
University of Alaska
Fairbanks, AK 99775

Arizona
University of Arizona
Tucson, AZ 85721

Arkansas
University of Arkansas
Little Rock, AR 72203

California
University of California
Davis, CA 95616

Colorado
Colorado State University
Fort Collins, CO 80523

Connecticut
University of Connecticut
Storrs, CT 06268

Delaware
University of Delaware
Newark, DE 19717

Florida
University of Florida
Gainesville, FL 32611

Georgia
University of Georgia
Athens, GA 30602

Hawaii
University of Hawaii
Honolulu, HI 96822

Idaho
University of Idaho
Caldwell, ID 83605

Illinois
University of Illinois
Urbana, IL 61801

Indiana
Purdue University
West Lafayette, IN 47907

Iowa
Iowa State University
Ames, IA 50011

Kansas
Kansas State University
Manhattan, KS 66506

Kentucky
University of Kentucky
Lexington, KY 40546

Louisiana
Louisiana State University
Baton Rouge, LA 70803

Maine
University of Maine
Orono, ME 04469

Maryland
University of Maryland
College Park, MD 20742

Massachusetts
University of Massachu-
  setts
Amherst, MA 01003

Michigan
Michigan State University
East Lansing, MI 48824

Minnesota
University of Minnesota
St. Paul, MN 55108

Mississippi
Mississippi State University
Mississippi State, MS 39762

Missouri
University of Missouri
Columbia, MO 65211

Montana
Montana State University
Bozeman, MT 59717

Nebraska
University of Nebraska
Lincoln, NE 68583

Nevada
University of Nevada
Reno, NV 89557

New Hampshire
University of New
  Hampshire
Durham, NH 03824

New Jersey
Rutgers State University
Cook College
New Brunswick, NJ 08903

New Mexico
New Mexico State University
Las Cruces, NM 88003

New York
Cornell University
Ithaca, NY 14853

North Carolina
North Carolina State
  University
Raleigh, NC 27695

North Dakota
North Dakota State
  University
Fargo, ND 58105

Ohio
Ohio State University
Columbus, OH 43210

Oklahoma
Oklahoma State University
Stillwater, OK 76078

Oregon
Oregon State University
Corvallis, OR 97331

Pennsylvania
Pennsylvania State
  University
University Park, PA 16802

Puerto Rico
University of Puerto Rico
Mayaguez, PR 00708

Rhode Island
University of Rhode Island
Kingston, RI 02881

South Carolina
Clemson University
Clemson, SC 29634

South Dakota
South Dakota State
  University
Brookings, SD 57007

Tennessee
University of Tennessee
Knoxville, TN 37901

Texas
Texas A&M University
College Station, TX 77843

Utah
Utah State University
Logan, UT 84322

Vermont
University of Vermont
Burlington, VT 05405

Virginia
Virginia Polytechnic
  Institute
Blacksburg, VA 24061

Washington
Washington State University
Pullman, WA 99164

West Virginia
West Virginia University
Morgantown, WV 26506

Wisconsin
University of Wisconsin
Madison, WI 53706

Wyoming
University of Wyoming
Laramie, WY 82071

District of Columbia
USDA, Room 3334, South
  Building
Washington, DC 20250

Canada

Address your letter to
  "Poultry Specialist":

Alberta
Department of Animal
  Science
University of Alberta
Edmonton, Alberta I6G
  2P5 Canada

British Columbia
Animal Science Depart-
  ment
University of British
  Columbia
Vancouver, British Colum-
  bia V6I 2A2 Canada

Manitoba
Department of Animal
  Science
University of Manitoba
Winnipeg, Manitoba R3I
  2N2 Canada

Nova Scotia
Department of Animal
  Science
Novia Scotia Agriculture
  College
Truro, Nova Scotia B2N
  5E3 Canada

Ontario
Department of Animal and
  Poultry Science
University of Guelph
Guelph, Ontario N1G
  2WI Canada

Quebec
Department of Animal
  Science
Macdonald College,
  McGrill University
Ste. Anne de Bellevue,
Quebec H9X IC0 Canada

Saskatchewan
Department of Animal and
  Poultry Science
University of
  Saskatchewan
Saskatoon, Saskatchewan
S7N 0W0 Canada

Poultry Suppliers

Culter's Supply
1940 Old 51
Applegate, MI 48401

Inman Hatcheries
Box 616
Aberdeen, SD 57402

**Keipper Cooping
  Company**
Box 249
Big Bend, WI 53103

Lynn Electric Company
2765 Main Street
Chula Vista, CA 91911

**Murray McMurray
  Hatchery**
Webster City, IA 50595

**Northern Wholesale
  Veterinary Supply**
Box 2256
Rockford, IL 61131

**Stromberg's Chicks and
  Game Birds Unlimited**
Box 400
Pine River, MN 56474

Organizations

**American Bantam
  Association**
c/o Eleanor Vinhage
Box 127
Augusta, NJ 07822

American Egg Board
1460 Renaissance Drive
Park Ridge, IL 60068

**American Minor Breeds
  Conservancy**
Box 477
Pittsboro, NC 27312

**American Poultry
  Association**
c/o Nona Shearer
26363 S. Tucker Road
Estacada, OR 97023

National Broiler Council
Box 28158 Central Station
Washington, D.C. 20005

National Egg Art Guild
c/o Lona Dagutis
924 Old Bethlehem Pike
Colmar, PA 18915

**International Chicken
  Flying Association**
Bob Evans Farm
Route 35
Rio Grande, OH 45674

Rare Breeds Conservancy
General Delivery
Marmora, Ontario Canada
KOK 2MO

**Society for the
  Preservation of Poultry
  Antiquities**
c/o Marion Nash
Box 102
Murphysboro, IL 62966

Further Reading

American Standard of Perfection (available from American Poultry Association, 26363 S. Tucker Road, Estacada, OR 97023)

Avian Bowl (study packet available from Clemson University, Bulletin Room 82, Poole Agricultural Building, Clemson, SC 29634)

Bantam Standard (available from American Bantam Association, Box 127, Augusta, NJ 07822)

Chicken Diseases by F.P. Jeffrey (available from American Bantam Association, Box 127, Augusta, NJ 07822)

A Chick Hatches, by Joanna Cole and Jerome Wexler (Morrow Junior Books). Out of print – check your library for a copy.

Eggcyclopedia (available from American Egg Board, 1460 Renaissance Drive, Park Ridge, IL 60068)

Egger's Journal (available from Crondall Cottage, Highclere, Newbury, Berks RG15 9PH, Great Britain, or c/o Elaine Nelisse, 11067 Madison Road, Montville, OH 44064)

"Information Bank" (available from American Bantam Association, Box 127, Augusta, NJ 07822; pamphlets on showing and breeding)

Poultry Press, Box 542, Connersville, IN 47331

Films and Videotapes

Chick Embryo Explanation
Thorne Films, Inc.
1707 Hillside Road
Boulder, CO 80302 (film)

Feeding the Flock
Agricultural Information
 Department, Drawer 3AI
New Mexico State
 University
Las Cruces, NM 88003
(video)

The Growing Embryo
Film Library
New York State College of
 Agriculture
Cornell University
Ithaca, NY 14850 (video)

*Introduction of Flock
 Health*
Central Mailing Services
Oklahoma State
 University
Stillwater, OK 74078
(video)

Chick, Chick, Chick
Film Library
Purdue University
West Lafayette, IN 47905
(film)

Glossary

This glossary contains two different groups of words. One group includes sounds chickens make. The other group includes words humans use when they talk about chickens.

Chicken Sounds

If you listen carefully, you will hear your chickens use specific sounds to communicate with each other and with you. The best way to learn chicken "language" is to pay attention when your birds engage in different activities. Soon you will be able to shut your eyes and know exactly what your chickens are doing by the sounds they make.

Chickens use at least thirty different sounds. How many can you identify? Don't be discouraged if at first you recognize only one or two. Learning a foreign language takes time.

Since sounds can be difficult to describe, collect chicken sounds on a cassette recorder. Use the numbers on the recorder's counter to keep track of each sound. In a notebook,

write each number and describe what the chicken was doing when it made that sound.

Hints for learning new sounds:

- What does a chicken say when it is ready to roost for the night?

- What does it say when a cat or other small animal comes near?

- What does a mother hen say when she finds something tasty for her chicks to eat? When does a rooster make the same sound?

To help you start your list, here are descriptions of some common sounds chickens make.

Brrrr. When a mother hen senses danger, she says "brrrr." Her chicks hear the warning and flatten themselves against the ground. They stay there until their mother says "cluck," telling them it is safe to get up.

Cackle. When a hen lays an egg she may cackle for several minutes after leaving the nest. You might think she is bragging, but she is really protecting her future offspring by drawing the attention of potential predators to herself as she moves away from the nest.

Cluck. A mother hen goes around all day saying, "Cluck... cluck... cluck...." She does this so her chicks won't get lost. They always know where their mother is by the sound of her voice. The loud, sharp sound is so distinct that a hen with chicks is often called a clucker or a cluck. Sometimes people call each other "dumb cluck," which is an insult to chickens.

Crow. Roosters crow for lots of reasons — to establish territory, to brag after winning a fight, and to keep the flock together. When a rooster is about to crow, he flaps his wings and stretches his neck. Nearby roosters who hear the crowing may answer. Occasionally, an old hen will crow if there are no roosters around.

Growl. When a hen is on the nest and you reach beneath her to collect eggs, she may fluff up her feathers and growl. She is telling you to leave the eggs alone. She is trying to protect the eggs so she can hatch them.

Hawk. If a bird flies overhead, one of your chickens (usually the highest-ranking one) will say something that sounds like "hawk." (Of course, they aren't really saying the word "hawk." It just sounds that way to humans.) The other chickens run for cover, in case it really is a hungry hawk looking for dinner. Chickens are easily fooled and sometimes holler "hawk" when it is only a butterfly. That is how the story of Chicken Little got started.

Peep. Chicks say "peep" even before they break out of their shells. After they hatch, chicks continue saying "peep" to keep in touch with each other and with their mother. When they are happy, their peeps are soft and infrequent. When they are unhappy, they peep louder and more often. Chicks have lots of different ways to say "peep." Listen carefully and you can tell when a peeping chick sees something scary, gets lost, or is ready to go to sleep.

Screech. If a dog grabs one of your chickens by the neck, or you surprise a chicken by suddenly picking it up, it lets out a shrill screech. Right away, a rooster or a senior hen will come to the rescue. Watch out! Even if your rooster is friendly, he may attack when he hears this sound.

Singing. Contented hens make a melodious sound that can only be described as singing. When you hear it, you will know you have done a good job raising your chickens, and you will feel happy, too.

Human Words

Humans use a lot of strange words when they talk about chickens. Some of the same words apply to other birds and animals as well. You will sound knowledgeable if you use these words when you talk to other people about your chickens. Add to this list as you learn new chicken-related words.

abdomen (n.). The lower part of a chicken's body between its pubic bones and its keel.

abdominal capacity (n.). The distance between the pubic bones and the keel bone, and between the two pubic bones.

abdominal depth (n.). The distance between the pubic bones and the keel bone.

abdominal width (n.). The distance between the two pubic bones.

as hatched (adj.). Description of chicks that have not been sorted by sex.

aviary netting (n.). Fencing woven in a honeycomb pattern with ½-inch openings.

bantam (n.). A miniature chicken that is about one-fourth the size of a regular chicken.

banty (n.). An affectionate word for bantam.

barbicels (n.). Tiny hooks that hold a feather's web together.

barnyard chicken (n.). A chicken of mixed breed.

barny (n.). An affectionate word for a barnyard chicken.

beak (n.). The hard, protruding portion of a bird's mouth consisting of two parts: an upper beak and a lower beak.

beard (n.). The feathers bunched under the beaks of such breeds as Antwerp Belgian, Faverolle, and Houdan.

bedding (n.). Straw, wood shavings, shredded paper, or anything else covering the floor of a chicken coop to absorb moisture and manure; also called *litter*.

billing out (v.). Using the beak to pull feed over the lip of a feed trough.

bleaching (v.). The disappearance of color from the beak, shanks, and vent of a yellow-skinned laying hen.

breed (n. and v.). A group of chickens that are like each other and different from other chickens; to pair a rooster and hen for the purpose of obtaining fertile eggs.

breeders (n.). Mature chickens from which fertile eggs are collected; also, people who manage such chickens.

brood (v. and n.). To care for a batch of chicks; the chicks themselves.

brooder (n.). A mechanical device used to imitate the warmth and protection a mother hen gives her chicks.

broody hen (n.). A hen that is ready to set.

broiler (n.). Young, tender meat chicken; also called a fryer.

candle (v.). To examine the contents of an egg with a light.

candler (n.). The light used to examine the contents of an egg.

cannibalism (n.). A bad habit some chickens have of picking each other's flesh or feathers.

cape (n.). The narrow feathers between a chicken's neck and back.

chalazae (n.). Two white cords on each side of a yolk that keep the yolk properly positioned within the egg white.

chick-chain (n.). A system for acquiring chicks with the promise of returning a few grown chickens.

chicken wire (n.). Fencing woven in a honeycomb pattern with 1-inch openings.

class (n.). A group of chickens competing against each other at a show.

classification (n.). The grouping of purebred chickens according to their place of origin, such as "American" or "Asiatic."

clean-legged (adj.). A description of breeds without feathers growing down their shanks.

clutch (n.). A batch of eggs that are hatched together, either in a nest or in an incubator; also called a *setting*.

coccidiosis (n.). A parasitic infestation that kills chicks kept in damp, unclean housing.

coccidiostat (n.). A drug used to keep chickens from getting coccidiosis.

cock (n.). A male chicken; also called a *rooster*.

cockerel (n.). A male chicken under 1 year old.

comb (n.). The red crown on top of a chicken's head.

condition (n.). A chicken's state of health and cleanliness.

conformation (n.). A chicken's body structure.

coop (n.). The place where chickens live; also called a *hen house;* at shows, the cage in which a chicken is exhibited.

crest (n.). A ball of feathers on the heads of breeds such as Houdan, Silkie, or Polish; also called a *top knot.*

crop (n.). A pouch at the base of a chicken's neck that bulges after the bird has eaten.

crossbreed (n.). The offspring of a hen and rooster of two different breeds.

cull (v. and n.). To eliminate a nonproductive or inferior chicken from a flock; the nonproductive or inferior chicken that is eliminated from the flock.

defect (n.). An imperfection that causes a chicken to lose points or to be disqualified from a show.

disinfectant (n.). Anything used to destroy disease-causing organisms.

disqualification (n.). A defect or deformity serious enough to bar a chicken from a show.

domestic (adj.). Not living in the wild.

down (n.). The soft, furlike fluff covering a newly hatched chick; the fluffy part near the bottom of a feather.

dressed (adj.). Cleaned in preparation for eating.

droppings (n.). Chicken manure.

dual-purpose (adj.). Term for chickens that are raised both for egg production and meat production.

dub (v.). To trim the combs, wattles, and ear lobes of Modern Game and Old English Game cocks.

egg tooth (n.). A horny cap on a chick's upper beak that helps the chick pip through the shell.

embryo (n.). A fertilized egg at any stage of development prior to hatching.

exhibition fowl (n.). Chickens kept and shown for their beauty rather than their ability to lay eggs or produce meat.

feather-legged (adj.). A description of those breeds with feathers growing down their shanks.

fertile (adj.). Description of an egg that is capable of producing a chick as a result of the joining of the rooster's sperm with the hen's ova.

flock (n.). A group of chickens living together.

fowl (n.). Domesticated birds raised for food.

free choice (adv.). Available at all times.

free-range (v. or adj.). To allow chickens to run loose so they can forage for food.

frizzle (n.). Feathers that don't lay flat, but instead curl so the chicken looks like it had a permanent.

fryer (n.). A tender young meat chicken; also called a broiler.

gizzard (n.). An organ that contains grit for grinding up the grain a chicken eats.

grit (n.). Sand and small pebbles eaten by a chicken and used by its gizzard to grind up grain.

hackles (n.). A rooster's cape feathers.

hatch (v. and n.). The process by which a chick comes out of the egg; a group of chicks that come out of their shells at roughly the same time.

hatchery (n.). A place where eggs are incubated and chicks are hatched.

hatching egg (n.). A fertile egg stored in a way that does not destroy its ability to hatch.

hen (n.). A female chicken.

hen-feathered (adj.). The characteristic of a rooster having rounded rather than pointed sex feathers.

hopper (n.). A container that is filled from the top and dispenses from the bottom, used for providing free-choice supplements and grit.

hybrid (n.). The offspring of a hen and rooster of different breeds, each of which might also be a crossbreed.

incubation (n.). The process of maintaining favorable conditions for hatching fertile eggs.

incubator (n.). A mechanical device for hatching fertile eggs.

infertile (adj.). The characteristic of an egg whereby it contains no sperm and therefore cannot produce a chick.

keel (n.). The breast bone, which resembles the keel of a boat.

lay ration (n.). Feed for laying hens.

litter (n.). Bedding.

mate (v. and n.). To pair a rooster with one or more hens; a hen or rooster so paired.

membrane (n.). A thin, soft, pliable layer.

molt (v.) To shed and renew feathers.

mortality (n.). Death due to disease or accident.

nest (n.). A secluded place where a hen feels it is safe to leave her eggs.

ova (n.). Female germ cells that become eggs.

oviduct (n.). The tube inside a hen through which a fully formed egg travels when it is ready to be laid.

parasite (n.). An organism that lives on or inside a host animal and derives food or protection from the host without giving anything in return.

pen (n.). A group of chickens entered into a show and judged together.

peck order (n. and v.). The social rank of chickens that establishes which one is boss, which is second in command, and so forth.

perch (n. and v.). The place where chickens sleep at night; the act of resting on a perch; also called *roost*.

pigmentation (n.). The color of a chicken's skin, beak, shanks, and vent.

pip (n. and v.). The hole a newly formed chick makes in its shell when it is ready to come out; also, the act of making the hole.

pin feathers (n.). The tips of new feathers coming through a chicken's skin.

plumage (n.). The total set of feathers covering a chicken.

poultry (n.). Chickens and other domesticated birds raised for food.

poultry netting (n.). Chicken wire, sometimes made of plastic.

predator (n.). An animal that hunts other animals for food.

pubic bones (n.). Two sharp, slender bones that end in front of the vent.

pullet (n.). A female chicken under one year old.

purebred (n. and adj.). The offspring of a hen and rooster of the same breed.

range-fed (adj.). Description of chickens allowed to run loose so they can forage for food; also called *free-ranged*.

ration (n.). A combination of ingredients that supply all of a chicken's dietary needs.

roaster (n.). A cockerel or pullet, usually weighing 4 to 6 pounds, that can be cooked in the oven.

roost (n. and v.). The place where chickens spend the night; the act of resting on a roost; also called *perch*.

rooster (n.). A male chicken; also called a *cock*.

saddle (n.). The part of a rooster's back just before the tail, usually covered with long pointy feathers.

scales (n.). The small, hard overlapping plates that cover a chicken's shanks and toes.

scratch (v. and n.). The habit chickens have of scraping their claws against the ground to dig up tasty things to eat; also, any grain fed to chickens.

set (v.). To keep eggs warm so they will hatch.

setting (n.). A group of hatching eggs in an incubator or under a hen.

sex feather (n.). A hackle, saddle, or tail feather that is rounded in a hen but usually pointed in a rooster (except in breeds that are hen-feathered).

sexed (adj.). Separated into pullets and cockerels.

sex-linked (adj.). A quality of newly hatched chicks making them easily sorted by sex.

shank (n.). The part of a chicken's leg between the claw and the first joint.

sickles (n.). The long, curved tail feathers of some roosters.

spent (adj.). No longer laying well.

spurs (n.). The sharp points on a rooster's shanks.

standard-bred (adj.). Conforming to the breed description in the *American Standard of Perfection* or the *Bantam Standard*.

started pullets (n.). Young female chickens that are nearly old enough to lay.

starter ration (n.). A ration for newly hatched chicks.

strain (n.). A flock of related chickens bred by one person for long enough that the offspring are uniform in appearance or production.

straight run (adj.). Newly hatched chicks that have not been sexed; also called *as-hatched*.

stub (n.). Down on the shank or toe of a clean-legged chicken.

trio (n.). A cock and two hens or a cockerel and two pullets of the same breed and variety.

type (n.). The size and shape of a chicken that tells you what breed it is.

typy (adj.). True to type.

unsexed (adj.). Not separated into pullets and cockerels; also called *as-hatched* or *straight-run*.

variety (n.). Subdivision of a breed according to color, comb type, beard, or feathering.

vent (n.). The opening through which a chicken emits eggs and droppings.

wattles (n.). The two red or purplish flaps of flesh that dangle under a chicken's chin.

web (n.). The network of interlocking parts that give a feather its smooth appearance.

zoning laws (n.). Laws regulating or restricting the use of land for a particular purpose, such as raising chickens.

INDEX

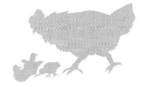